Digital Signal Processing Technology: Essentials of the Communications Revolution

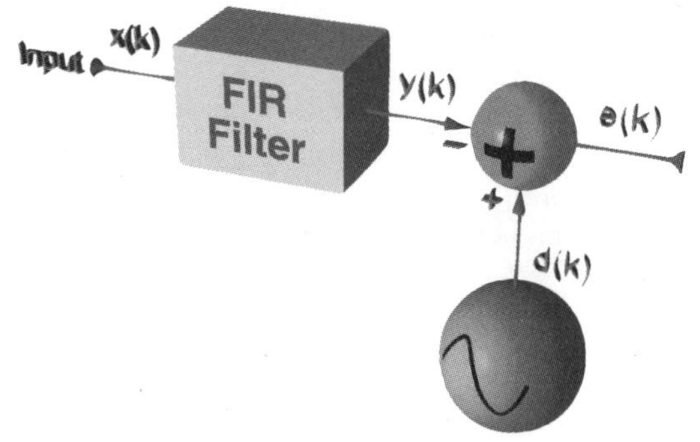

Doug Smith,KF6DX

Production

Michelle Bloom, WB1ENT—Production Supervisor
Paul Lappen—Composition and layout
Sue Fagan—Cover design
David Pingree, Michael Daniels—Technical illustrations
Jayne Pratt Lovelace—Proofreader

Newington, CT 06111-1494
ARRLWeb: www.arrl.org

Printed in USA

First Edition
Second Printing, 2003

ISBN: 0-87259-819-5

Dedication

To my lovely wife, Dari, who graciously endured unmentionable hardships during the production of this book: My dear, your wisdom and support made it possible.

About the Author

Doug Smith, KF6DX, has over 20 years experience designing communications systems and circuits for commercial, military and Amateur Radio applications. His computer programming career began on mainframes like the CDC 7000 about eight years before the IBM PC was first introduced, when punch cards and nine-track tape drives were prevalent.

Doug's areas of concentration have covered a wide range of systems, including control systems for radio, automatic link establishment, frequency synthesis and DSP. He was involved in the design of some innovative products for amateurs, including the Kachina 505DSP, the first full-power, computer-controlled HF amateur transceiver. His current technical work extends to digital voice-coding methods.

Since 1998, Doug has edited *QEX/Communications Quarterly*: Forum for Communications Experimenters, published bimonthly by the ARRL (**www.arrl.org/qex**). He was the recipient of the 1998 ARRL, Doug DeMaw, W1FB, Technical Excellence Award, and remains involved in various League activities.

On the air, Doug enjoys rag-chewing and RTTY operation. Away from the shack, he is an avid amateur astronomer and photographer. He enjoys teaching children about the joys of Amateur Radio. Doug may be contacted through League headquarters: Editor, QEX, 225 Main St, Newington, CT, 06111.

Contents

Preface

A digital revolution has already transformed telecommunications and virtually every other facet of our lives. We stand at a point where hardware capabilities have very nearly caught up with the development of theory; this situation has only evolved within the last decade or so of the 20th century. Demands for more communication bandwidth have driven DSP execution speeds higher. Lately, awesome computing power has become available at reasonable cost. It is somewhat ironic that those higher speeds are forcing computer designers to become RF engineers, since their PCB traces are UHF transmission lines; some RF engineers now focus entirely on DSP—we have exchanged our roles!

It's just as well, because when we look into DSP, we find that a good grasp of both digital and analog concepts is important. At the interface between the two, many trade-offs present themselves. As you read these pages, therefore, don't be too surprised to find a fair discussion of analog system requirements. That doesn't mean that DSP need be overly complex to learn: it just involves acquiring a new set of skills.

Perhaps it is appropriate to define what we mean when we say DSP. In this volume, DSP refers not only to the process of numerically manipulating sampled signals, but also to how those signals are used. For example, the output of a speech-processing circuit may not be a processed analog signal, but a control signal that performs another function. Also, the input to a DSP algorithm may not be a sampled analog signal. That distinction is not particularly emphasized where it arises, but may be important to keep in mind.

Also, note that every DSP construct described in this book has an analog equivalent that, although perhaps incredibly implausible to build, could function nearly the same. We've used analog-circuit equivalents where possible to aid understanding. For example, the fast Fourier transform—DSP's best spectrum-analysis method—may be thought of as a bank of band-pass filters and detectors. That wonderful equivalence exposes a rich variety of applications, some of which are described. I'm confident another crop will have emerged as soon as the ink on this page has dried.

We hear much lamenting that radio design is too difficult for the amateur these days, since everything has gotten so complex. Make no mistake: a smaller percentage of designers are working on state-of-the-art development than before. Those few, however, are actually making it easier for the rest of us! Highly integrated sub-systems are now readily available to the experimenter, offering features above and beyond anything that was out there even five years ago. I guess there is no going back now that we've tapped into technology that gives us higher performance at lower cost.

It's important that technology—including DSP—serve one purpose: to make things better. Science has a different end: to discover what makes things tick. While this may be a fine line to draw, we must state that the material in this book is presented from an engineering perspective. We've made an effort to balance the theoretical with the practical. Mathematics has been used freely where it presents itself as the most efficient form of expression. Elsewhere, plain English seems to do well enough.

Chapter 1 provides a brief history of DSP and an overview of its application—not only in communications, but also in exploration for oil, restoration of recordings, astronomy and other fields. General benefits and drawbacks of the technology are discussed. Chapter 2 examines the sampling theorem, aliasing and certain mechanisms at play in real data converters. The relationship between bandwidth and sampling rate is central. Methods of changing the sampling rate of an already-sampled signal are discussed and reasons for wanting to do this are explained. Representation of signals is a major consideration in DSP systems. Chapter 3 takes up this subject to make clear how numbers are actually stored and manipulated.

Chapter 4 begins a look at DSP algorithms with perhaps the most important subset: digital filters. It covers the construction of well-known types and their properties. Adaptive filtering is covered in a later chapter. Chapter 5 introduces the concept of analytic signals and their representation as complex numbers. We revised the chapter in this printing to correct details of certain digital modulation modes. Chapter 6 examines digital coding methods for speech, including error detection and correction in digital transmission. Chapter 7 examines digital frequency synthesis methods, including direct digital synthesis (DDS), fractional-N, and hybrid techniques.

In Chapter 8, duality between time-domain and frequency-domain representations of signals is discussed. It begins with adaptive-filtering algorithms. The chapter continues with a treatment of Fourier transforms and their inverses. The conclusion looks at variations of Fourier-transform methods and their application.

Chapter 9 takes us through the design of digital transceivers at the block-diagram level. The discourse starts with DSP at AF stages, then moves the digitization point closer to the antenna in steps. Receivers and transmitters are considered separately but commonality of circuits is illustrated. Chapter 10 describes DSP hardware: general-purpose and dedicated DSPs, data converters and DDC chip sets. Chapter 11 discusses software aspects of DSP design.

For journeyman and advanced readers, Chapter 12 introduces some exciting areas of current research. I believe many of these shall find their way into Amateur Radio soon.

This book begins with basic concepts and gradually brings in more complex ideas. The discussion of sampling theory requires no special math skills; the remainder assumes a working knowledge of algebra, trigonometry and binary arithmetic. Experience with engineering statistics (summation signs) is helpful. Projects require the ability to program a computer and, for advanced development, the purchase of a DSP evaluation kit and other off-the-shelf hardware and software. I hope the book will help you reach a better understanding of this rapidly developing topic.

Acknowledgment

Many thanks to the kind folks at ARRL who participated in the making of this book, including Mark Wilson, K1RO; Joel Kleinman, N1BKE; Jan Carman, K5MA; Dave Sumner, K1ZZ; and to the Production crew in Newington for making it shine. Dennis Silage, K3DS, of Temple University has my eternal appreciation for reviewing the material and giving me a good sanity check. His input was invaluable.

Doug Smith, KF6DX
Seymour, Tennessee
October, 2003

Introduction to DSP

DSP stands as one of the greatest innovations of the last millennium. Because of rapid advances in microprocessor technology, DSP systems are today revolutionizing the way we live our lives—even, in many cases, without our awareness of them. This chapter takes a look at the role of DSP in fields other than communications and includes a bit about its history.

DSP Without Computers

In the centuries before digital computers were invented, all calculations had to be performed by hand. In the case of celestial mechanics, for example, these calculations became quite complex; large teams of mathematicians would work for weeks on end to obtain a single numerical result. The accomplishments of Galileo Galilei (1564-1642) and Johannes Kepler (1571-1630) are particularly astonishing given that they lacked the calculus of Newton.

In the late 1500s, a Scot named John Napier (1550-1617) discovered a thing called the logarithm that simplified arithmetic by replacing multiplication with addition.[1] Two numbers may be multiplied by adding their logarithms and finding the anti-logarithm of the sum. He started to compile a gigantic book of logarithms while Tycho Brahe (1546-1601) waited in vain for the thing that would speed his calculations manifold. Napier (and Brahe) expired before the book was finished, but Napier's friend Henry Briggs (1561-1630) completed the work and published it in London. Briggs' logarithms were a boon to both earthly and celestial navigators.

It was not until the late 1600s that Nicolaus Mercator (1619-1687) found that the natural logarithm of a number could be represented as the sum of certain fractions relating to that number. This type of sum is now called a *McLaurin* series, even though Mercator and James Gregory (1638-1675) were working with them before McLaurin's birth. It is perhaps ironic that history remembers the name Mercator for maps (Gerardus Mercator, 1512-1594, not Nicolaus); Colin McLaurin (1698-1746) himself invented an important rule for solving systems of equations that is today called Cramer's rule!

At any rate, the discovery of the McLaurin series for natural logarithms meant that very accurate values could be computed, rather than retrieved from a book, for any number, and with reasonably few terms. It is an example of a summation of discrete terms that converges to a result. We shall see this type of construct again later.

Mathematicians of the time realized that breaking down calculations into smaller, more easily soluble parts was their first line of attack on many problems. Iterative methods were also common: the repetitive computation of a set of equations that converges on a result. Most of these techniques are today unnecessary and forgotten; but a few of them are still in use. Their inventors hardly would have called them DSP algorithms, but they remain as part of DSP's legacy.

After Isaac Newton (1642-1727) sprang his genius on the mathematical world, myriad applications for calculus revealed themselves. Often, it was necessary to integrate curves or surfaces for which equations were either unknown or intractable. Practitioners found they could get a reasonable approximation by breaking the curve into many short segments and treating the segments as straight lines or sections of a parabola; the areas under these segments were then added to get the final result. While this method was known long before his birth, Thomas Simpson (1720-1761) mentioned it in his widely read book and thus it is remembered as *Simpson's Rule*. Again, history gets it wrong as one of the greatest mathematicians of 18th-century England is remembered not for his unique contributions, but for something that wasn't really his at all.

Simpson's Rule is nevertheless a very early example of the representation of continuous waveforms as discrete samples. In fact, it is remarkable how similar the process is to that of digital-filter algorithms. You may note that similarity when you get to Chapter 4.

Using the calculus, Joseph Fourier (1768-1830) discovered the relationship between application of heat to a solid body and its propagation.[2] By 1812, he had a complete explanation of the effect and was awarded a great prize for scientific accomplishment—the judges were Laplace, Legendre and Lagrange! Fourier could scarcely have imagined the impact his work has had on almost every imaginable discipline. His methods as applied to analysis of signals are treated in Chapter 8.

When working with *Fourier transforms*, a great many integrals are involved; those often include integration over infinite spaces of time and distance. Folks discovered, though, that there was little point in including the contributions of a heat source that had been applied, say, several hours before. Equilibrium had already been reached. We may suppose that at some stage, a bright engineer asked himself, "Why should I integrate over infinity? I will just integrate the significant (recent) parts and *truncate* the input data to that length." Combined with integration using Simpson's Rule, this approach constitutes *finite-impulse, discrete Fourier transforms* that are quite similar to what is done in modern DSP systems. It shows that discrete signal processing was not foreign to those who came long before us. Although it provided a way to efficient computing, they did not have the computers to make it shine.

Fourier transforms became so important in physics that many highly skilled mathematicians applied their wits to breaking down their computational complexity. There is evidence that Carl Gauss (1777-1855) worked on the problem with some success, even anticipating Fourier in many ways.[3] At the start of the 20th century, Carl Runge (1856-1927) analyzed the problem intensively and produced a solution very similar to today's *fast Fourier transform (FFT)*. Because even the reduced calculations were not practical by hand, the discovery was largely overlooked until Cooley and Tukey picked up the gauntlet in the 1960s.[4] By then, digital computers were ready for the task.

It is shown that the roots of DSP run deep. The development of the atomic bomb, with Richard Feynman (1918-1988) in charge of computations, could not have been accomplished without it. The orbital mechanics necessary to put Armstrong and Aldrin on the Moon would have been nearly impossible. While those endeavors had computational assistance, the help was pretty basic compared with what we have today.

DSP With Computers

In general, modern DSP systems characterize and modify analog signals, producing other analog signals as their outputs. Note that this is not always the case, though; the output of a DSP circuit might just as well be the opening of a squelch gate or the triggering of a VOX. Alternatively, DSP algorithms may use non-linear stimuli as their inputs and even produce non-linear outputs; but for the most part, we are interested in exploiting DSP's advantages in linear systems we're used to.

Most DSP hardware systems take a very flexible approach in that they include means of translating analog signals to digital form, a microprocessor to manipulate those digital signals, and then of converting the results back to analog. Such a system may be defined completely by the program running on the microprocessor, and is therefore *software-defined*. Both hardware and software may have some say-so about how signals are converted between analog and digital formats. Setting of the sampling rate and analog input bandwidth may be placed under microprocessor control. Most DSP systems operate at fixed sampling rates and ratios, but certain advantages may come to the designer who makes them variable.

Many DSP systems look like analog, two-port networks: analog signals into and out of a black box that somehow transforms the signals. That transform may be virtually anything you can think of that is physically possible, such as a filter or a speech processor. It may also be some esoteric function like a bandwidth compressor: Programming determines function. So a large part of DSP involves programming of digital computers; however, advances in circuit integration are producing dedicated DSP chips that perform specific functions very well, such as filtering and mixing. As against that, though, field-programmable logic devices make excellent platforms for dedicated DSP functions. We thereby retain the full measure of flexibility.

DSP systems are popular because they tend to exhibit better performance than analog equivalents. One reason for this is illustrated in filter design. An analog filter consists of certain physical components, such as inductors and capacitors, that determine its characteristics. To meet tight tolerances on its passband or stopband, such a filter demands equally tight tolerances on its component values. In a digital filter, the "component" values are stored as numbers in a computer: They are known as *coefficients*. These coefficients are the same from unit to unit and so, very nearly, is the frequency response of the filters. Temperature variations are nonexistent. This freedom from variation lets us attempt complex filters that would be beyond reason in the analog world. Digital filters having more than one hundred coefficients are commonly used. Imagine trying to construct an analog filter with half that many poles! Another positive note is that DSP filters don't exhibit higher loss as they get more complex—but they do get an additional noise contribution. This isn't usually a significant factor, though, in practical circuits.

A second outstanding reason for using DSP is that it usually eliminates a lot of hardware from traditional analog designs, thus lowering cost. An expensive set of analog band-pass filters, for example, might be replaced with a much larger set of superior digital filters—as many as the associated coefficient memory would hold—at little additional cost. Modulator, demodulator, squelch and speech-processing circuits might be eliminated entirely in favor of DSP. As hardware capabilities progress, the digitization point in transceivers will move closer to the antenna jack, eliminating still more hardware. This trend comes at a price, though, for DSP hardware; that price is somehow proportional to the bandwidth in which we're interested. DSP also tends to tax time resources because of the need to learn programming constructs.

DSP in Other Technical Fields

DSP has found its way into fields other than communications. In many cases, advancement of the technology has been driven forward by necessity; in other cases, it has served mainly an analytic role. This distinction may be illustrated with examples from the engineering world, the arts, the pure sciences and medicine. The following examples are offered as evidence of DSP's versatility and emphasize its wide-ranging impact on humankind.

DSP in the Search for Fossil Fuels

DSP is a marvelous tool that has helped unlock some of nature's best-kept secrets. It has achieved what was generally considered quite implausible even

10 years ago. Many algorithms developed for one particular application have found new uses in otherwise unrelated areas. An outstanding example of this is the concept of *echo-cancellation* as used in the exploration for oil and other underground fossil fuels. Algorithms nearly identical to those used in geophysics were originally created to eliminate echoes on the public switched telephone network. Much of this early work was done at Bell Laboratories. Similar algorithms are now employed to combat multi-path distortion on radio communications paths.

Come back in time with us to 1939. Suppose you are a geophysical engineer for an oil company and your job is to build a machine that finds pockets of "black gold" underground. The boss wants you to get cracking, because WWII has just started and need for the stuff is about to soar. He knows his current teams are achieving only a 1-in-12 strike rate—that is, they find significant oil deposits at only 8.3% of their test wells. You have already worked out a clever scheme to improve on this and filed a patent application for it, but it will be another three years before the patent is issued. Your name is R. T. Cloud.[5]

You reason that to find something that is hidden from view, you must apply some kind of stimulus to it and see how it reacts. Your weapon of choice is a hydraulic ram. To test your invention, you locate it where you already know a large subterranean oil pocket exists. You arrange for the ram to deliver swept sine-wave excitation to the surface of the Earth. Then you bury some microphones at strategic spots (geophones) to pick up sounds coming back from the Earth. Ordinarily, it is very quiet down there because Texas does not get many earthquakes; but when you fire the ram, you get a lot of sound—seemingly from everywhere!

Refer to **Fig 1.1**. The hydraulic ram applies impulses to the Earth and these impulses propagate away in all directions, including straight to geophone #1, which is located very close to the ram so as to capture only the incident wave. Distant geophones pick up waves directly through the Earth through paths near the surface; these are of little interest. They also record impulses reflected from objects and from the discontinuities between layers of dissimilar material deep beneath the surface. You know this because of the delayed times of arrival of these echoes. Armed with a reasonable guess about the properties of the Earth's crust beneath you (and perhaps some core samples), you may induce the distance traveled by the echoes and therefore the depth of the objects or reflecting layers.

By placing enough geophones, you obtain enough *spatial diversity* to triangulate the exact locations of reflecting matter underground. You compare the data from the various geophones and find data from each that correlate with the others. Further reasoning reveals that you could build a circuit using a tapped delay line and attenuators that would allow you to cancel any particular echo by manually adjusting the taps and the attenuation applied to each. See **Fig 1.2**. You build it and it works. When you examine the final delay taps and attenuator settings, you find they form an image of the propagation properties of the path taken by that particular echo. You go into the field and discover you can

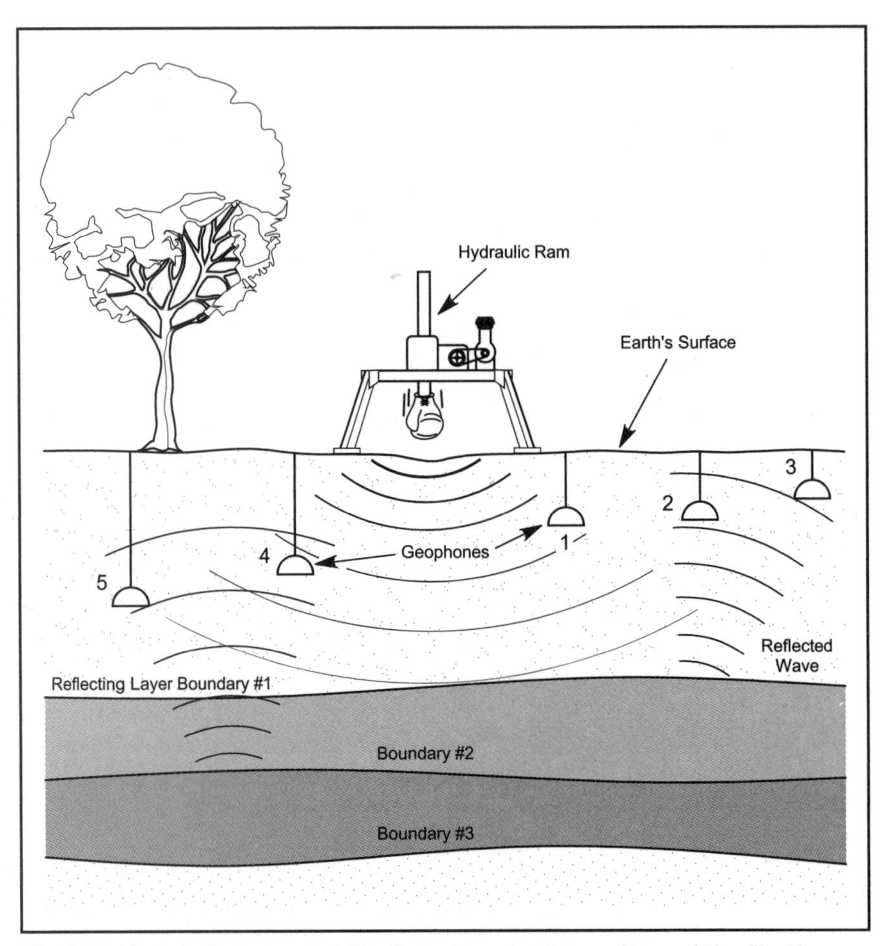

Fig 1.1—Hydraulic ram applying impulses to the surface of the Earth.

tell the boss just where to drill. Your strike rate is now 1 in 6—you have doubled output and you get a big raise!

Now another bright geologist comes along and points out that your wells are having to go deeper than ever before to hit home; most of the easy oil has already been found. He reasons that the Earth's crust is thinner at the ocean floor, so it is a better place to look, but sinking the hydraulic ram to the sea floor is a bit much. He decides that a single impulse is as good as swept sine waves (they are both broadband) and he packs some explosives in his kit and heads to sea with your machine. Some say that he is already "at sea" with that idea, but they are wrong. His more powerful arrangement requires some revamping of the machine, but eventually his strike rate reaches 1 in 4.

As we zoom forward to the present day, geophysical engineers are able to

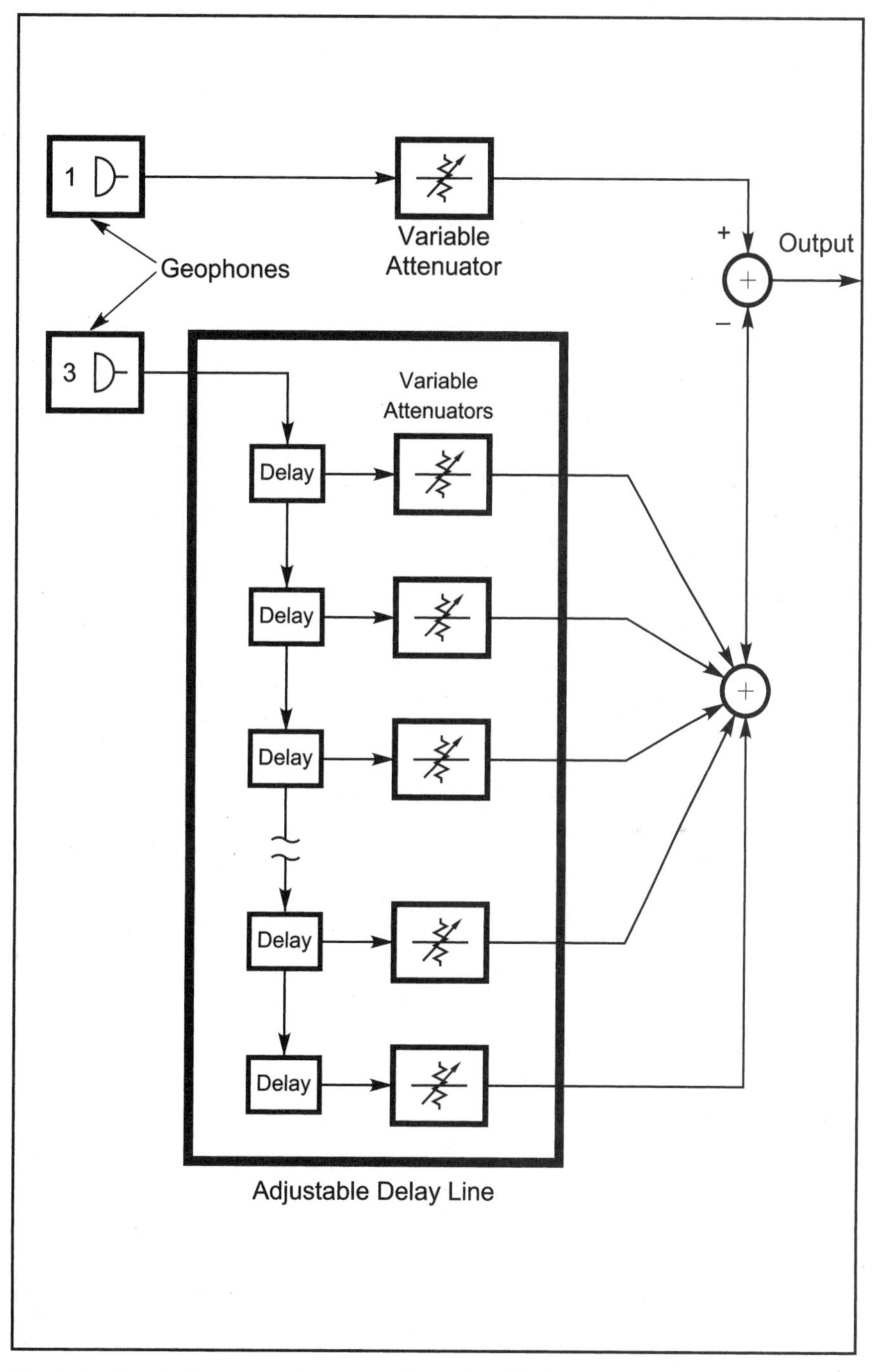

Fig 1.2—Manual system for canceling a particular echo.

produce 3-D renderings of subsurface structures using techniques that grew from this early work. They can even plot the motions of oil in pockets and discern its temperature and density. Adaptive DSP methods have thus risen to dominance in the field of oil exploration.

DSP in the Recording Studio

Now suppose you are a top-flight recording engineer and your company embarks on a project to restore some old recordings. One set includes Enrico Caruso (1873-1921) singing in Milan in the early part of the last century. The original recording engineer was Emil Berliner (1851-1929). The audio has since been transferred from its original medium, wax discs, to vinyl discs, but you are not sure of the quality of equipment that was used to do that. You do know a lot about the original recording equipment and, of course, you know that the original recording hall has not changed much over the years.

The recordings are plagued by pops, clicks and broadband background noise that are obviously undesirable. They are also colored by the inadequacies of Berliner's equipment. You don't want to eliminate too many of the echoes, since they constitute much of the acoustic quality of the room, but you do want to compensate for the poor directional characteristics of those ancient microphones. You wish to fix these things so the result will sound like a modern, digital recording—a tall order, to be sure!

With exact measurements of the hall and of some of the old equipment, you determine a net *transfer characteristic* for the system, and you arrange to build a DSP system that corrects the original system's flaws through the process of *deconvolution*, discussed in detail later. What you cannot precisely define, though, is when a pop or a hiss will appear on the recording; however, you do know a pop when you hear it and this information may be used to eliminate it. Being a smart engineer, you know that you must take what the game gives you and use all the information at your disposal to achieve the goal. You realize you are fortunate in that you do not have to process the data in "real time" and you may take as long as you like; but you also know that to manually eliminate all the faults including the background noise would be a herculean task. You therefore draw on DSP algorithms to do it for you.

One critical bit of information in your possession is that the pops, clicks and noise do not resemble the desired program material very much. Transient events on the recording do not sound like Enrico at all, and thus ought to be removable; broadband noise does not sound much like him either. His output is characterized by a tremendous range of tonal qualities and by a fairly large dynamic range. The key word here is tonal, since the singer produces waves that are the sums of large numbers of sinusoids at different frequencies and their harmonics. These waves are therefore repetitive in some way over short time spans. This fact may be used to differentiate between desired and undesired content.

So, very much like geophysicist Cloud, you build a circuit that allows you to correlate current chunks of audio with recent chunks. You accept only the current chunks that resemble the recent ones, and reject the rest. The pops and clicks are eliminated, since they do not repeat themselves over short time frames. Broadband noise also does not correlate with itself, since it is random. This *adaptive noise-reduction* technique will be covered in detail in Chapter 8.

DSP at the Telescope

DSP may also be applied with advantage to the field of image compression and enhancement. In fact, this is one of the fastest-growing areas of current research. As in the above examples, DSP is extended to two and three spatial dimensions by virtue of a particular system's architecture. In the science of astronomy, men and women study the part of the universe that lies beyond Earth's atmosphere, notwithstanding that meteors cause a bit of overlap. A large part of their work involves extracting information from data taken near the limits of our ability to measure. In many cases, a very small number of photons arriving at the film or sensor plane may be all that is required to confirm or refute some premise. DSP is handy in these and many other situations related to image processing.

Thanks to the rapid proliferation of charge-coupled devices (CCDs) and personal computers, many folks have been exposed to digital image processing. Computer programs are now readily available that perform very sophisticated transformations on digital image data: A group of techniques employing *maximum-entropy* methods is among the foremost of those. Still, basic properties of an image are familiar terms to any television viewer: brightness, contrast, hue, resolution.

By digitizing image data, control over all these properties is given to DSP, with advantage. For example, many astronomical objects of interest are high-contrast, but some are not. The surfaces of planets and moons within the Solar System, faint nebulae and galaxies may reveal unseen parts of their structures to the DSP-equipped astronomer. Image traits such as contrast and hue may be manipulated to enhance available information content as long as the original traits are not destroyed. As will be noted later, even resolution may be improved by DSP control of data acquisition. One of the first applications of CCDs was in observing the surface of the Earth from orbit. This government project fostered the development of enhancement techniques that many enjoy today on their home PCs. Many types of semiconductor sensors are now employed, including CMOS, to record images.

That does not mean that film is obsolete, though. Astronomers have long held that film-plane images with resolution in the 10-μm range are the best obtainable. This is about equal to the resolution of CCDs, but very large CCDs are difficult to build because of the limitations of current silicon technology. Millions of picture elements (pixels) require millions of transistors on a single slice of material and at least a few of them are bound to fail on fabrication or in the field. Multiplexing techniques have minimized this problem, especially in large LCD displays. Astronomers can live with a few bad pixels, just as they lived with dust specks and scratches for so many years. Image scale may be very important in recording all the information from a particular subject and, at the time of this writing, film still wins by its sheer number of pixels. An 8″×10″ piece of film, for example, contains a number of 10-μm pixels:

$$\left(80\ \text{in}^2\right)\left(645\times10^6\ \mu\text{m}^2/\text{in}^2\right)\left(0.01\ \text{pixels}/\mu\text{m}^2\right)=516\times10^6\ \text{pixels} \tag{1}$$

Whether the data were digitized at acquisition or later, processing them all is a formidable task; but again, you do not necessarily have to do it in real time. Correlation properties mentioned above form the basis for deciding how image data differ from other types of signals and, therefore, how they should be differently processed. One continuing, robust segment of the image-processing community is focused on data compression. Various standards have emerged in this category, including JPEG, MPEG, *wavelet transforms* and fractal coders. Descriptions of these are mainly outside the goals of this book. Let it suffice that achievements in image compression have exceeded expectations to the point where digital television (DTV) broadcasting is plausible; however, we have to acknowledge that advances in modulation formats are equally responsible.

As mentioned, DSP is also used in astronomy to enhance data *as they are gathered*. This is often a requirement for best performance. Telescopes have been built that employ adaptive reflecting elements whose alignment changes rapidly in response to variations in the atmosphere, temperature, and so forth to keep an image steady. The trend these days is to take advantage of whatever spatial diversity a particular system gives you by using smart DSP algorithms. Even without this, though, DSP may be used to cancel distortions encountered in image formation. Witness the design of a "contact lens" that was placed in the optical path of the Hubble Space Telescope to correct its "vision." Scientists and engineers found the refractive function that nearly nullified the gross error in the original grinding of the primary mirror—it would have been useless without modern DSP ray tracing.

DSP in Medicine

Medical doctors have learned some DSP, too. As in the case of echo cancellation above, physicians may employ DSP to cancel a pregnant woman's heartbeat and listen to the heartbeat of an unborn child within her.[6] Again, the spatial diversity of the sensor (stethoscope) array may be exploited to accomplish what otherwise would be impractical.

One stethoscope is placed near the mother's heart and produces a signal that contains little of the heartbeat from the fetus. Another is placed farther down the abdomen and produces a signal that contains both heartbeats. A DSP system is constructed that finds the correlation in the two signals—that which they have in common: the mother's heartbeat. The system is made to perfectly cancel the mother's heartbeat and what is left is that of the fetus. As you will see below, many other criteria may be used to condition signals for use in DSP. Some use spatial diversity, some use bandwidth and some use temporal qualities to differentiate between desired and undesired signals. The possibilities are endless. Before we dive into advanced topics, though, a discussion of how analog signals are sampled is in order.

Digital Sampling

Fundamental Sampling

Sampling may be simply described as the process of making periodic measurements of a signal and storing them. For purposes of this discussion, we define a signal as something in the physical world that can be measured and that contains information. It would be nice if those measurements captured all the information contained in a signal, but that is not always the case. Frequently we are interested in only one signal of many. We may discuss taking samples of signal parameters such as bandwidth or signal-to-noise ratio (SNR), just as well as amplitude versus time. Also, instead of taking samples at regular intervals of time, we may elect to take them at regular intervals in space; or, we may even talk about intervals of time-space and time-frequency. In electronics, sampling most often refers to measuring a signal's voltage at regular time intervals. This is the simplest case and most other representations stem from it.

This process is illustrated in **Fig 2.1**. A sine wave is shown being sampled at times denoted by a sampling function. Note that the frequency of the sine wave being sampled is much less than the *sampling frequency,* f_s. That is, we are taking many samples during each cycle of the sine wave. Note that the sequence of samples still resembles a sine wave. Although it does not contain information about the actual voltage between samples, we may state that all the information about amplitude versus time has been acquired by the sampled signal. This sampled signal also contains new information, though: information about the sampling process.

Difference in information content would be evident were we to examine the spectra of the two signals. All the energy in the sine wave is concentrated very near a single frequency; the sampled signal's spectrum is obviously not the same, since it is composed of steps separated by the sampling period. The

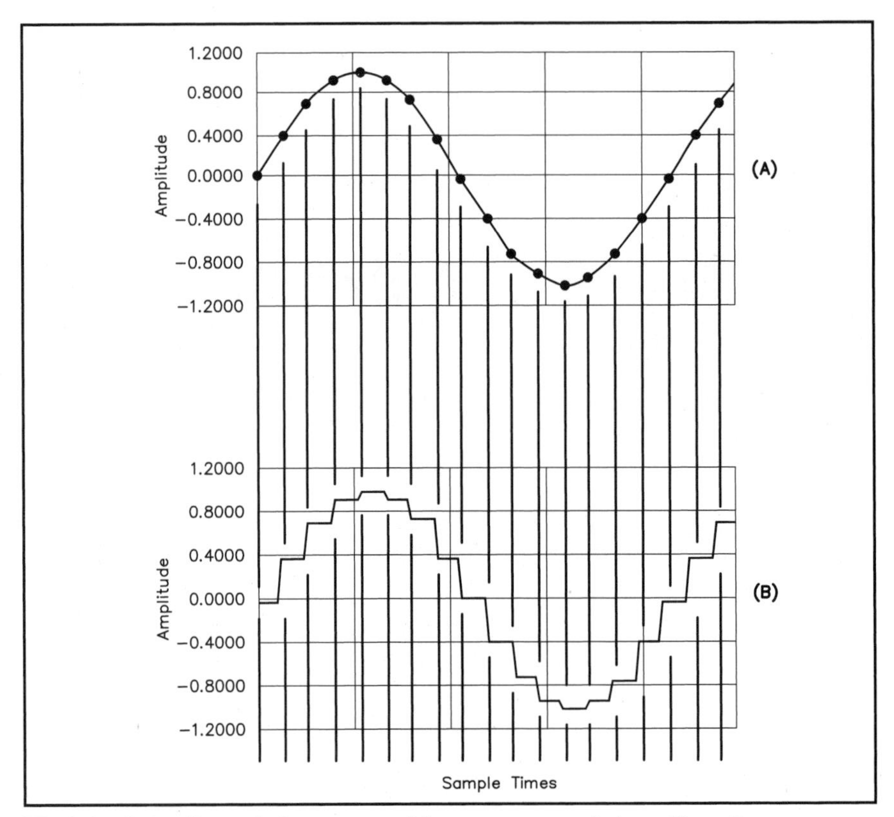

Fig 2.1—Sampling of sine wave of frequency much less than the sampling frequency.

sine wave's spectrum is shown in **Fig 2.2A** above the spectrum of the sampling function in Fig 2.2B. The sampled signal is just the *product* of the sine wave and the sampling function; its spectrum is the *convolution* of the two input spectra, as shown in Fig 2.2C.[7] This shows that sampling may behave as a mixing process: They each perform a multiplication of two input signals.

Note that the sampled spectrum repeats at intervals of f_s, the sampling frequency. These repetitions are called *aliases* and are as real as the fundamental in the sampled signal. Each contains all the information necessary to fully describe the original sine wave. What is evident in Fig 2.2C is easily shown: The first alias appears at frequency $f_s - f_{sine}$; the second, at $f_s + f_{sine}$; the third, at $2f_s - f_{sine}$; the fourth, at $2f_s + f_{sine}$, and so on. Energy in these aliases declines with increasing order, since the harmonic energy in the sampling function declines with increasing harmonic number. Fourier analysis, discussed in

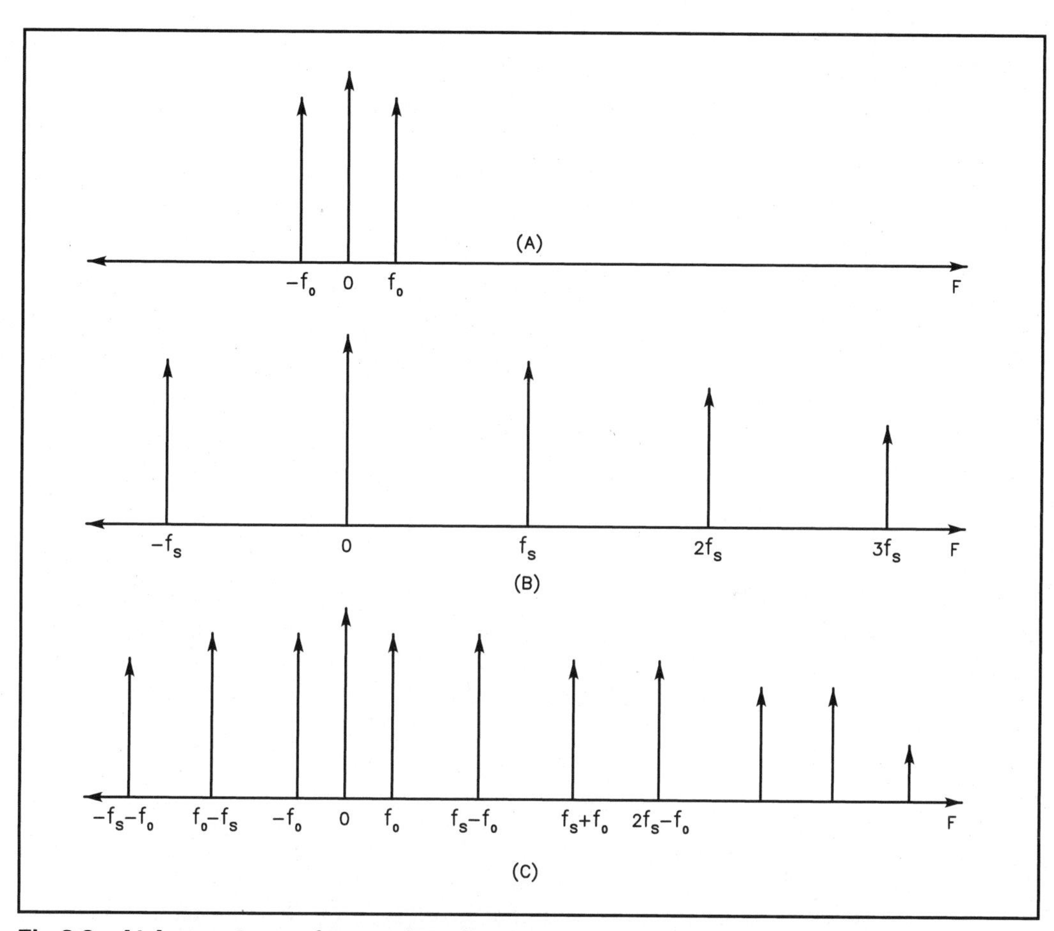

Fig 2.2—At A, spectrum of an analog sine wave. At B, spectrum of the sampling function. At C, spectrum of the sampled sine wave.

detail later, reveals that alias amplitudes are equal to:

$$A_n = A_1 \left\{ \frac{\sin\left[\pi\left(n \pm \frac{f}{f_s} - 1\right)\right]}{\pi\left(n \pm \frac{f}{f_s} - 1\right)} \right\} \tag{2}$$

where n is the harmonic number and f is the input frequency of the signal being sampled.

Another obvious feature of Fig 2.2 is that were we to increase input frequency, f, pairs of aliases would approach each other at $f_s/2$, $3f_s/2$, and so forth. When $f = f_s/2$, the output would be a square wave having amplitude proportional to the sine of the input phase. When $f > f_s/2$, the first alias and the fundamental exchange their places in halves of bandwidth f_s. A sine wave of frequency $f_s - f$ would produce a sampled signal identical to that of a sine wave of frequency f, as shown in Fig 2.2.

This demonstrates that to avoid aliasing, we must limit the bandwidth of our sampler's input to half the sampling frequency: That sampling rate is the often-misquoted *Nyquist frequency*. An analog filter is typically employed to do this and is called an *anti-aliasing* filter. Once aliasing has been incurred, nothing may remedy it and information about input signals may be destroyed. Higher-frequency signals take on the identities of lower-frequency signals, and vice versa. This frequency-translation property may be exploited, though, to minimize sampling frequencies. As described below, this is quite desirable in many instances.

Harmonic Sampling

Let us look at the case where the sampling frequency is less than that of an input sine wave; that is, $f > f_s$. See **Fig 2.3**. Notice that the shape of the sampled signal no longer matches the input signal; it retains the shape of a sine wave, but of lower frequency. This is the situation, mentioned above, that is ordinarily to be avoided; but a downward frequency translation is useful in the design of radio transceivers. It is equivalent to a frequency translation obtained in a mixer, which analogy was mentioned before. With certain restrictions, it allows sampling of signals that are much higher than the sampling frequency. In practical systems, higher sampling frequencies place a higher burden on sampling hardware to achieve accuracy, and on software to complete processing tasks between samples. They also usually incur a significant current-consumption penalty. Any technique that minimizes these factors is therefore quite desirable.

Caution is required, though, since an input signal near twice the sampling frequency would produce the same sampled signal as that of Fig 2.3. To use this technique, then, we must first bandwidth-limit the input, as before; but this time, a band-pass anti-aliasing filter (BPF) is called for. This technique is called *harmonic sampling*.

Now the largest bandwidth for the filter is $f_s/2$. Placing the filter's pass-band between the fundamental (or some harmonic) of f_s and the point half way to the next higher harmonic ensures that the entire bandwidth is usable. A frequency translation will take place equal to an integral multiple of f_s, but no information about the input signals will be lost. A spectral representation of harmonic sampling is shown in **Fig 2.4**. Again, the sampled spectrum is the convolution of the two input spectra, as in the mixer analogy above. No in-band aliasing occurs because the sampled bandwidth is less than half f_s.

Harmonic sampling is also called *bandpass sampling*. Its use is critical in

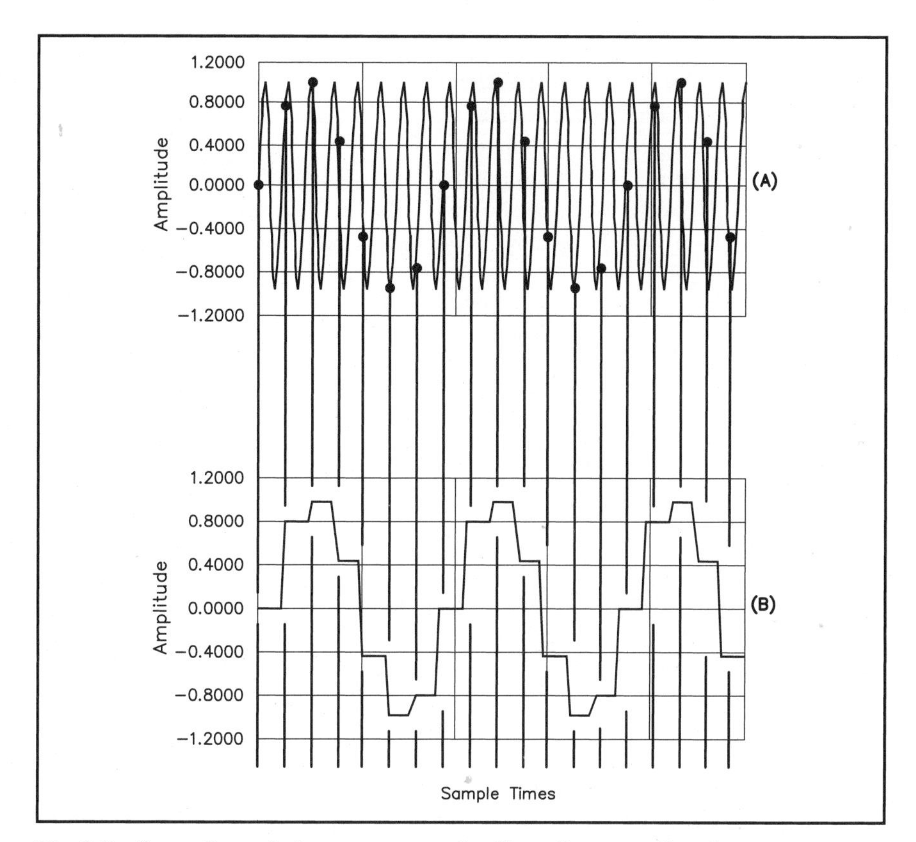

Fig 2.3—Sampling of sine wave greater than the sampling frequency.

the design of IF-DSP and digital direct-conversion transceivers.[8] The subject is introduced again with detail as we explore digital transceiver architectures in Chapter 9.

Data Converters and Quantization Noise

The device that performs sampling is generally called an *analog-to-digital converter (ADC)*. At each measurement, an ideal ADC produces a number that is directly and exactly proportional to input amplitude. Computer representations of this number mean that only a finite number of values are possible; usually, the number is in straight base-two or binary format having some number of bits, b, available. An 8-bit ADC, for example, can only give one of 256 values. This means the amplitude reported is never exact, but only the closest of those at hand. The difference between the actual input value and the reported is called the *quantization error*.

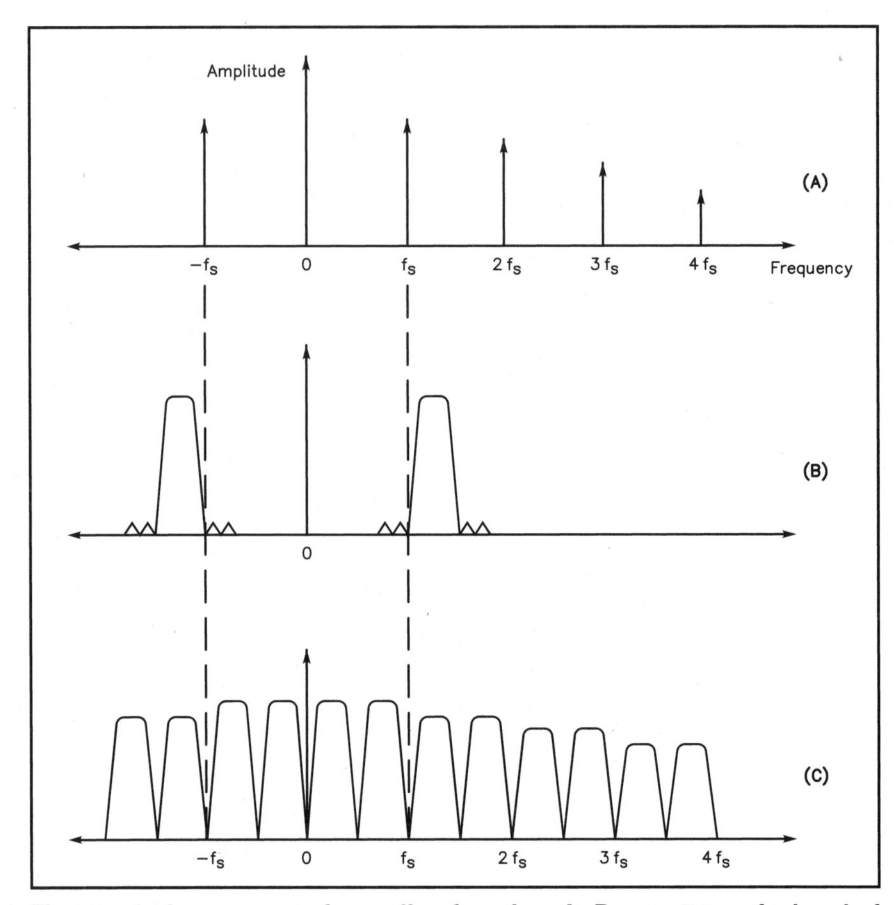

Fig 2.4—At A, spectrum of sampling function. At B, spectrum of a band of real signals. At C, spectrum of a harmonically sampled band of real signals.

Assuming that the input signal is changing and covers a fairly large range of quantization values, the error is just as likely to be positive as negative. It is also just as likely to be small as large, within certain limits. Hence, this error signal—a sequence in its own right—is pseudo-random and appears as *quantization noise*. In a perfect ADC, the error cannot exceed $\pm^1/_2$ of the least-significant bit of the converter; therefore, this is the error signal's peak amplitude. The sequence reported by the ADC can thus be thought of as the sum of the real input signal and the quantization noise.

Quantization noise is normally spread uniformly over the entire sampling bandwidth of $f_s/2$. From the previous discussion, we might expect that its amplitude is somehow proportional to the characteristics of the input signal. We want to find how it limits our maximum SNR in data converters. That maximum is very likely to occur when the input signal occupies the entire range of quantization levels, since its power will be maximized.

Let us call this "rail-to-rail" voltage V_{max}. Remember that this is the peak-to-peak input voltage. The smallest voltage step the converter can resolve is then:

$$\Delta V = \frac{V_{max}}{2^b} \tag{3}$$

We have stated that the error signal, e, has a range of $-\Delta V/2 \le e \le \Delta V/2$, or a peak-to-peak amplitude of ΔV. In our first use of engineering statistics, the RMS noise power into a 1 Ω load is found by integrating the square of the error (power is proportional to voltage squared) over the range of possible errors, each value of which is as likely as the next:

$$\sigma^2_{noise} = \frac{1}{\Delta V}\int_{-\frac{\Delta V}{2}}^{\frac{\Delta V}{2}} e^2\, de$$

$$= \frac{1}{\Delta V}\left[\frac{e^3}{3}\right]_{-\frac{\Delta V}{2}}^{\frac{\Delta V}{2}} \tag{4}$$

$$= \frac{1}{\Delta V}\left[\frac{\Delta V^3}{24} + \frac{\Delta V^3}{24}\right]$$

$$= \frac{\Delta V^2}{12}$$

The RMS noise voltage is just the square root of this, or:

$$V_{noise} = \frac{\Delta V}{\sqrt{12}} \tag{5}$$

When this is applied over the input bandwidth of $f_s/2$, the noise power per unit bandwidth (per Hz) is:

$$\frac{\sigma^2_{noise}}{\frac{f_s}{2}} = \frac{\Delta V^2}{12\left(\frac{f_s}{2}\right)} = \frac{\Delta V^2}{6f_s} \tag{6}$$

in watts/hertz. The power into a load of R ohms is just this divided by R. It is, perhaps surprisingly, not related to the nature of the input signal but only to the sampling frequency. We stated the input signal was large when we started. Small signals that do not exercise many quantization levels are liable to produce different results. In fact, when the input signal and f_s have a harmonic relationship, quantization effects tend to concentrate the noise at discrete frequencies. This may have an impact on *dynamic range*, as discussed further below.

Now for the SNR calculation. The RMS input power of a V_{max} sine wave, again normalized to one ohm, is:

$$P_{sine} = \frac{\left(2^b \Delta V\right)^2}{8} \qquad (7)$$

and so the SNR is found by taking the ratio of this value to the noise power:

$$\begin{aligned} SNR &= \frac{\dfrac{\left(2^b \Delta V\right)^2}{8}}{\dfrac{\Delta V^2}{12}} = \left(\frac{3}{2}\right)2^{2b} \\ &\approx 6.02b + 1.76 \text{ dB} \end{aligned} \qquad (8)$$

This equation conveniently places an upper limit on the dynamic range of a b-bit ADC. Additional pseudo-random noise sources appear in ADCs, though, that tend to limit performance still further.

Aperture Jitter

Noise is introduced in ADC results by variations in the exact times of sampling. While this effect is described here in a negative light, later it will be shown that with control, this noise may actually be helpful in extending dynamic range.

Phase noise or jitter in an ADC's clock source, as well as other inaccuracies in sampling mechanisms, may produce undesired phase modulation of the sampled signal. Again assuming the effect is uncorrelated with the input signal, this *aperture-jitter noise* will be uniformly distributed across the input BW. We may express the mean aperture jitter as σ_a in seconds: This is the given quantity and we want to extract the noise power produced by the phase modulation it causes. Now in this case, it is obvious that the input signal's frequency will come into the equation because the modulation index will change. In addition, it clearly matters how large σ_a is with respect to the sampling period, $1/f_s$.

It is now convenient to switch to discussing frequencies in both radians per second (angular format) and in hertz. Recall the simple relation:

$$\omega = 2\pi f \tag{9}$$

For small phase deviations, a PM signal may be represented by[9]:

$$\begin{aligned} V_{PM} &= A\cos\left(\omega_c t + \beta \sin \omega_m t\right) \\ &= A\left[\cos\omega_c t - \beta\left(\sin\omega_c t\right)\left(\sin\omega_m t\right)\right] \\ &= A\cos\omega_c t - \frac{A\beta\cos\left(\omega_c - \omega_m\right)t}{2} + \frac{A\beta\cos\left(\omega_c + \omega_m\right)t}{2} \end{aligned} \tag{10}$$

where ω_c is the carrier, A its amplitude, and ω_m is the modulation frequency, both in radians per second; β is the modulation index, or the peak phase deviation in radians (from Reference 9). The first term in Eq 10 is the carrier; the second, the lower sideband; and the third term is the upper sideband. For $\beta \ll 1$, theory shows that very little energy is contained in higher-order sidebands.

The phase deviation caused by the jitter is just equal to the ratio of jitter to carrier period, times 2π radians:

$$\beta = \frac{2\pi\sigma_a\sqrt{2}}{\left(\frac{2\pi}{\omega_c}\right)} = \omega_c\,\sigma_a\sqrt{2} \text{ radians} \tag{11}$$

Substituting this value into a sideband term in Eq 10 and assuming A = 1 produces a single-sideband noise power of:

$$P_{\text{SSB noise}} = \left(\frac{\beta}{2\sqrt{2}}\right)^2 = \frac{\omega_c^{\,2}\,\sigma_a^{\,2}}{4} \tag{12}$$

There are two sidebands, so the total noise power is twice that:

$$P_{\text{total noise}} = \frac{\omega_c^{\,2}\,\sigma_a^{\,2}}{2} \tag{13}$$

Now the maximum sine-wave input has A = 1 and its power is 1/2. Maximum SNR is therefore:

$$\text{SNR} = \frac{\left(\frac{1}{2}\right)}{\frac{\omega_c^{\,2}\,\sigma_a^{\,2}}{2}} = \omega_c^{\,-2}\,\sigma_a^{\,-2} \tag{14}$$

This equation reveals that values of β near 10^{-5} are where aperture jitter becomes a problem for 16-bit converters. For example, in the case where f_c = 100 kHz and σ_a = 20 ps, SNR ≈ 98 dB. Assuming uniform noise distribution to $f_s/2$, dividing Eq 13 by $f_s/2$ yields the noise density in W/Hz:

$$ND = \frac{\left(\frac{\omega_c^{\,2}\,\sigma_a^{\,2}}{2}\right)}{\left(\frac{f_s}{2}\right)} = \frac{4\pi^2 f_c^{\,2}\,\sigma_a^{\,2}}{f_s} \tag{15}$$

Note that this effect increases with the squares of the input frequency and the sampling jitter, but in inverse proportion to only the first power of the sampling frequency. That makes it difficult to maintain dynamic range while increasing sampling frequency. Jitter is a major concern for DDC designers dealing with VHF and UHF signals and very high sampling frequencies, such as for those working on high-speed data communications systems.

It is sometimes useful to think of the noise density-to-signal ratio for a particular converter. Divide Eq 15 by the maximum signal power of $^1/_2$ to obtain:

$$\frac{ND}{\left(\frac{1}{2}\right)} = \frac{8\pi^2 f_c^{\,2}\,\sigma_a^{\,2}}{f_s} \tag{16}$$

Nonlinearities

Nonlinearity in data converters means distortion and more noise. Quantization steps in any converter are not perfectly spaced and conversion results are contaminated by the inaccuracy. In general, two types of nonlinearity are characterized: differential nonlinearity (DNL) and integral nonlinearity (INL).

DNL is a measure of output nonuniformity from one input step to the next. It is expressed (in bits) as the maximum error in the output between adjacent input steps over the entire input range of the device. We're discussing the accuracy of the smallest steps a converter can resolve. Noisy, low-order IMD products produced by this effect tend to influence the dynamic range of any particular device.

Manufacturers have recently begun specifying *spurious-free dynamic range* (SFDR) for their devices under actual operating conditions—thank you! Usually, this is given for single-tone conditions; two-tone measurements are not as common. Obviously, nonlinearities in a data converter will result in IMD that may have bearing on various applications. In addition, a harmonic relationship between input signals and f_s may tend to concentrate spurious energy in discrete bands. That is a troublesome subject to illustrate with mathematics; a myriad of aliases and in-band distortions may limit performance unless careful attention is paid to specification of data converters. It is not hard, though, to show that jitter—as much as it is deplored above—may help dissipate those discrete spurs. That subject is treated in detail later.

Converters are considered *monotonic* if a steady increase in input signal always results in an increase in output. Backward steps may cause unexpected problems in systems working close to resolution limits. Manufacturers often offer different grades of converters that are specified to some number of least-significant bits: $\pm^1/_2$ bit at the minimum, for example, to maintain monotonicity.

INL is a measure of a device's large-signal-handling capability. To test it, we first inject a signal of amplitude A and take the output. Then, we inject a signal of amplitude 100A and compare the result with 100 times what we got before: We expect the output to increase in exact proportion. INL is a measure of the output error between *any two* input levels. Input vs output may be plotted and maximum deviation from a straight line is then easy to see. This effect produces additional harmonic distortion (HD) and IMD that may be quite undesirable. Typical values for INL center on ±1 bit.

Oversampling and Sigma-Delta Converters

Eqs 6 and 15 show that as f_s increases, noise density decreases in direct proportion. That means that if the sampling frequency were artificially increased by some large factor N, then the sampled signal digitally filtered to reduce its bandwidth by the same factor, a SNR improvement of nearly N would be obtained. Quantization and aperture-jitter noise would be spread over N times the bandwidth it would otherwise occupy; most of it would be removed by filtering. This technique is known as *oversampling*.

So-called *sigma-delta* converters use oversampling to achieve high SNRs. They employ single-bit quantizers at very high speed and digital filters to reduce bandwidth and sampling rate, thus obtaining noise reduction. These and other data converter types are further discussed in Chapter 10. Ways of reducing sampling rate in tandem with bandwidth are treated in a later section of this chapter.

Digital-to-Analog Converters: Additional Distortion Sources

Digital-to-analog converters (DACs) translate binary numbers back to analog voltages—the inverse operation of ADCs. They suffer from all the effects above, as well as few of their own. One unique distortion of DACs is one of frequency response: *sample-and-hold distortion*.

Typical DACs are sample-and-hold devices: They continue to output their most-recent value throughout the sample period. The result is a step-wise representation of output data that acts as a low-pass filter, albeit a mediocre one. The frequency response of such a filter is:

$$H_f = \frac{\sin\left(\frac{\pi f}{f_s}\right)}{\left(\frac{\pi f}{f_s}\right)} \tag{17}$$

Note the sin(x)/x form of this function, called a 'sinc' function. Its value is defined to be unity at f=0, where the function is otherwise discontinuous. The high-frequency roll-off is quite undesirable in many instances. For example, were the output frequency equal to $f_s/4$, an attenuation of about 1 dB would occur. Correction may be made for this, but increasing the sampling rate is often an easier solution. That is discussed more below.

When the output of a DAC changes from one voltage to another, it obviously cannot do so instantaneously; a finite time is required for the voltage to reach its new value. This is generally known as the *settling time*. It is usually defined as the time required to settle within some number of voltage-equivalent bits of the final value.

Glitch energy or *glitch area* may be defined as the product of the voltage error during settling and the settling time itself. While we know volt-seconds are not units of energy, we may assume that a DAC is driving some kind of load; thus, glitch area may be translated into units of energy (watt-seconds). Settling mechanisms are important factors in the production of spurious outputs in DACs. Manufacturers usually specify glitch area for their high-speed devices. It is an especially important specification for digital-oscillator applications. Those are discussed further in Chapter 7.

Sampling in More Than One Dimension

The above discussion concerns itself with sampling at regular intervals of time. Notice that although sampling of voltage is how DSPs get their information, those samples may well represent a variety of quantities, such as water flow, brightness or sound intensity. Those things may be sampled not only over time but over space, too.

As is often the case in recording, two microphones on a stage may be arranged to detect sounds at different places to help create a two-dimensional effect for the listener: stereo. That creates a *two-dimensional array* or *lattice* of data. More than two microphones may be used lying roughly in a single plane (the stage) and two spatial dimensions are involved. A pair of stereo loudspeakers is normally arranged in a single line, though, which has only one dimension; but the recording engineer has captured two sets of sound intensity data from two different places. The two sets of data correlate with one another based on the distance of any particular sound source from each microphone and its frequency. When the recording is replayed, a listener's brain detects the correlation and a two-dimensional image of what was recorded is restored.

The foregoing example produces an array of data that were sampled at regular intervals of both time and space. The data may be analyzed in each of those domains separately and each may be defined as a *fundamental domain* of the data array.[10] Spatial analysis lends itself to other problems, such as determining the direction of arrival of a radio signal. A set of omnidirectional antennas replaces the microphones above, and are again usually arranged in a line, or in a single plane. Assuming the distance between antennas is not large compared to the distance to the signal source, the output of each is identical to the next but for a small time delay. A signal may arrive at one antenna before it arrives at another because of its finite propagation speed, c. This fact may be used to steer the pattern of the antenna array to some extent.

Let us say that the signals from two antennas separated in space are simply added together to form the input to a receiver. Placing a fixed delay line in one of the antenna output leads will make the array most sensitive to signals arriving at some angle or angles to the line between the antennas. When the delay is zero, the angle is naturally 90°. Changing the delay alters the radiation pattern of the array.

Criteria for spatial analysis extend to other signal properties, as well. We may elect to analyze the strength of a signal received at each antenna separately over time. Especially with ionospheric propagation, it is often found that signal strengths vary markedly at each of the two antennas. When one antenna delivers a strong signal, the other's may be weak. That is because each antenna is really receiving many time-delayed copies of the source that have refracted from many different parts of the ionosphere. Path lengths vary, and so do times of arrival; at one antenna site, signals may cancel to produce fading while at the other site, they reinforce. This leads to a *spatial-diversity reception* system that switches antennas based on the received signal strength at each, maximizing energy into the receiver over time.

Because each antenna's signal may be sampled at regular intervals of time, Fourier transform techniques may be employed to analyze frequency content over time, resulting in terms of *time-frequency*. Adding spatial dimensions to the mix means that we can discuss quantities of *space-frequency*. These terms come into treatments of speech analysis and compression, and of adaptive antenna arrays. Chapter 12 contains further details of these and other esoteric modes.

Changing the Sampling Rate: Multi-Rate Processing

Lowering the Sampling Rate: Decimation

There are many reasons for wanting to change the sampling frequency of an already-sampled signal. Perhaps foremost among those is the desire to minimize the sampling frequency f_s for some particular signal of interest. That desire is driven by the need for more time between samples to perform calculations—the calculations necessary to do modulation, demodulation, squelch, and other functions vital to radio transceivers. Systems that employ *sampling-rate conversion* may be referred to as *multi-rate DSP* systems.

We may be faced, for example, with a situation where we have to filter a broadband input to extract a narrow-band signal. The broadband input must be limited to a bandwidth of half the initial sampling rate; as signals are digitally filtered to lesser bandwidth, sampling rate may be reduced. Reduction in sampling rate is achieved by resampling the already-sampled signal at a lower rate. This process is generally known as *decimation*; and the filter, as a *decimation filter*. Decimation is most often performed entirely in the digital domain on signals previously sampled at a higher rate.

Decimation itself is simple: Just discard samples according to the decimation ratio. For a *decimation ratio* of two, keep only every other sample; for three, keep only every third sample. Decimation is usually performed by such integer ratios, although it does not have to be. A useful application of noninteger decimation occurs when mating two systems with different sampling rates. Fractional decimation is discussed more later.

A decimation filter runs at the higher sampling rate, eliminating components above half the new, lower sampling frequency f_{s_new} that would cause aliasing after the rate reduction. Depending on the application, this filter may be a BPF, LPF, or HPF, so long as it limits bandwidth.

Perhaps the simplest case is that of a low-pass decimation filter ahead of the decimator. **Fig 2.5** illustrates such an arrangement. Normalized to a 1-Hz sampling rate, the ideal frequency response of the low-pass is a "brick wall" such that:

$$H_{\omega} = \begin{Bmatrix} 1, \omega \le \dfrac{\pi}{D} \\ 0, \text{otherwise} \end{Bmatrix} \tag{18}$$

where D is the decimation ratio. The output of the decimator is an alias-free signal having a sampling rate D times lower than that of the input. Although the filtering operation is linear and time-invariant, it may readily be shown that combining it with decimation yields a time-variant system (see Reference 7).

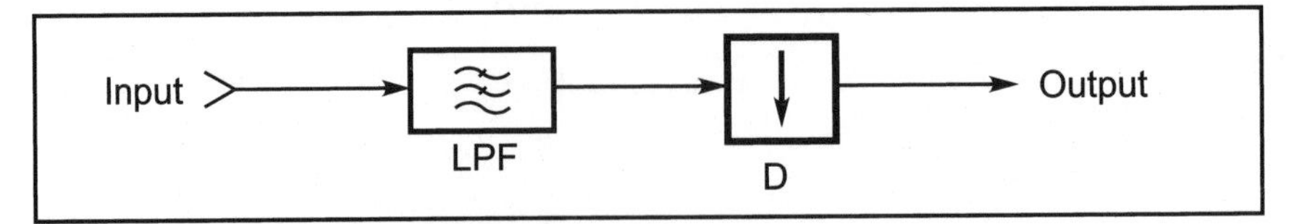

Fig 2.5—Decimation LPF and decimator.

Astute readers may be wondering, "During decimation, why compute filter outputs that must only be discarded? Isn't that a waste of processing time?" Well, yes, it is! Just calculate those samples that are to be kept. This is equivalent to running the decimation filter at the lower rate. So decimation reduces the sampling rate through filtering and it makes the filtering a bit easier, too.

Increasing the Sampling Rate: Interpolation

While low sampling frequencies are pleasant for the reasons mentioned, converting digital signals back to analog produces aliases that may not be far removed from the desired response. There, they are difficult to eliminate with the mandatory analog anti-aliasing filter. Sample-and-hold effects distort the frequency response much more, too. An artificial increase in sampling frequency is often called for. This is usually known as *interpolation*.

Again, this is typically done by integer factors. Samples with a value of zero are inserted between data samples to produce a longer sequence; these are then filtered at the higher sampling rate to remove aliases of the lower sampling rate. The filter is called an *interpolation filter* and is most often a low-pass, although high-pass and band-pass have been used to advantage. The ideal interpolation filter has the same type of brick-wall frequency response as the decimation filter of Eq 18; however, note that interpolation actually produces more usable bandwidth at its output than that at its input. There is room to fit other signals in this extra bandwidth, but an analog anti-aliasing filter is then more difficult to build.

Fractional Sampling-Rate Conversion

Adopting the notation of Proakis et al.,[11] let us consider sampling-rate conversion by a rational factor U/D. Perhaps the sampling rate of one subsystem is exactly $^3/_2$ of another. If they need to interface with one another, a fractional sampling-rate conversion is needed. The conversion may be made by first interpolating by U = 3, then decimating by D = 2. This cascade is shown in **Fig 2.6.** In this simplest form, the input bandwidth should not exceed 1/3 of the output bandwidth to avoid aliasing. Adding interpolation and decimation filters dodges this limitation. See **Fig 2.7**. Interpolation must occur first in the chain lest information in the input signal be lost. It cannot necessarily be done the other way round.

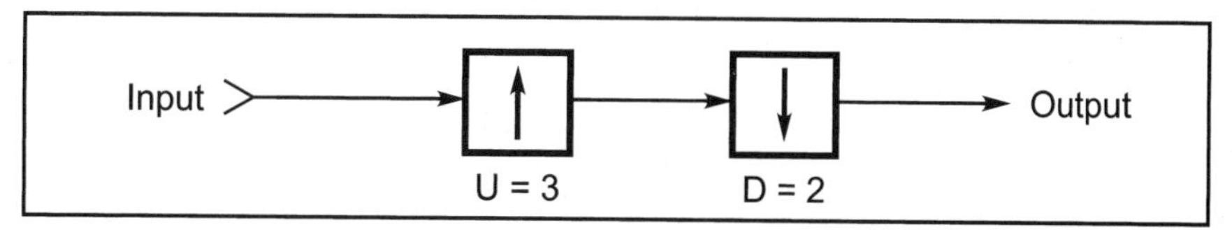

Fig 2.6—Fractional decimation by first interpolating, then decimating by integer ratios.

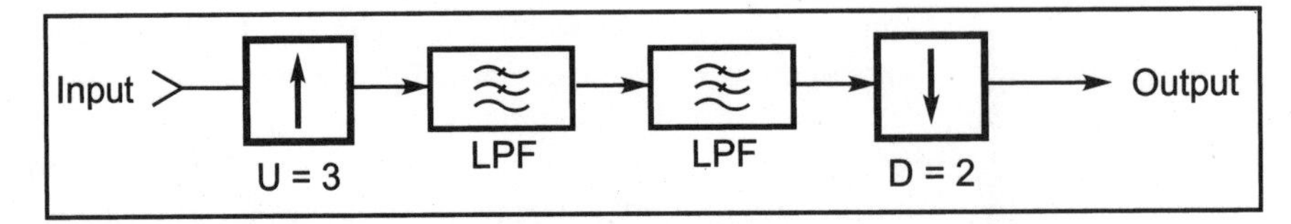

Fig 2.7—Fractional decimation with filtering.

Because of the way the filters in Fig 2.7 work, they may operate with economy at a common sampling rate, and thus may be combined into a single filter. They are, after all, linear, time-dependent processes that are chained together. DSP filtering is treated in Chapter 4; a mathematical derivation of how to combine the filters' *impulse responses* may be found there, as well.

Before we can consider the details of filtering operations necessary to continue this thread, a discussion of how samples (numbers) are represented in computers is in order. Accuracy of a DSP result is affected not only by imperfections in data converters, but also by computational errors; these, in turn, are caused by the limitations of binary-number representation as demonstrated in the following chapter.

Computer Representations of Data

Numerical Formats

This chapter introduces two popular binary number-storage formats and compares their attributes: fixed-point and floating-point. A hybrid notation, block floating-point, is also treated. Many DSP texts begin with the subject of numeric representation because it is so important to performance of actual systems. Accuracy and dynamic range issues are examined for each format. A knowledge of binary arithmetic is assumed.

Fixed-Point Format

In *fixed-point* format, the available number of bits, b, is usually used to represent a number whose absolute value is less than one; that is, the number is a *signed fraction* within the range ±1. Using typical ADCs and DACs, samples would normally be handled in this manner. Most-significant bit b is taken to be the *sign bit*. When the sign bit is zero, the fraction represented by the remaining b – 1 bits is positive; when one, the fraction is negative. Negative, fixed-point numbers are stored as the two's complement of their absolute value. The *two's complement* of a number may be found by subtracting it from zero.

The fraction is just a sample's value as compared with the maximum-possible amplitude. The *radix point*, or separation between integer and fractional parts, is assumed to reside left of the second most-significant bit (MSB). This is terribly convenient in DSP calculations because the product of two fractions less than unity is always another fraction less than unity. This advantage will become apparent.

Refer to **Fig 3.1**. The largest positive number is $2^{b-1} - 1$, and the most-negative number is -2^{b-1}. With b = 16, those numbers are equal to 0111 1111 $1111\ 1111_2$ and $1000\ 0000\ 0000\ 0000_2$, respectively. The dynamic range of

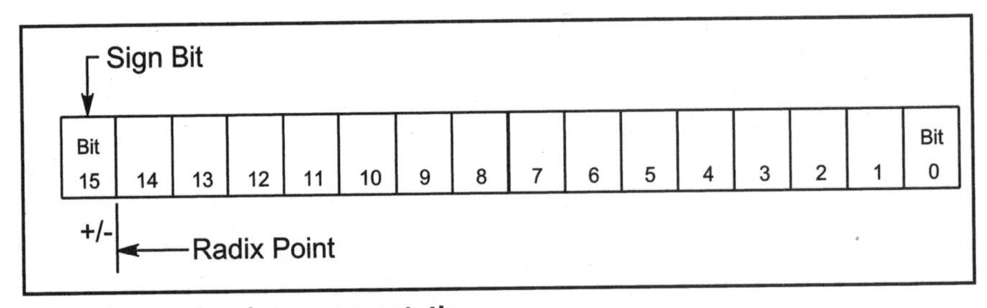

Fig 3.1—Fixed-point representation.

such representation (about 98 dB) is limited by quantization effects, just as in the case of data converters above. Numbers must be *rounded* or *truncated* to 16 bits. In fact, computational quantization noise is calculated in exactly the same fashion as for data converters in Chapter 2. It may not always be clear whether quantization noise from the two sources add, or not; poorly understood correlation effects often contribute to energy concentration at discrete frequencies. It is safe to write, though, that DSP-system dynamic range is near optimum when the bit-resolution of the data converters nearly matches that of the signal processor. Certain advantages are retained by selecting a processor having slightly more bit-resolution than the data converters, especially when filtering is involved. This is described further below.

While multiplying fixed-point numbers does not present a problem in numeric representation, adding or subtracting them does. Adding two fractions less than unity may obviously produce a result greater than unity: *overflow*. This leads to the need for scaling of data, alternate numeric representations, or both. DSP algorithms may be examined for computational dynamic range and input data scaled to prevent overflow; but in some cases, the range must be extended.

One way of extending the computational dynamic range of fixed-point representation is to add more bits. The extra bits may be used to represent the integer part of the number. Take a total number of bits, 2b, and place the radix point in the middle, instead of to the left of the MSB. See **Fig 3.2**. Now this number may be multiplied with another having the same representation in straight binary fashion to produce a result having 4b bits. The radix point in the result is placed at 2b bits. Many processors don't have registers with enough bits to do this directly, though. When b = 16, 4b = 64 and that is a large result. An interesting situation occurs when the data-handling capacity of the digital signal processor is only b bits. The integer and fractional parts are handled separately in the computation, much as real and imaginary parts are in complex mathematics.

Adopting the notation c.d, where c is the b-bit integer part and d the b-bit fractional part, two numbers c.d and e.f may be added at once in binary fashion:

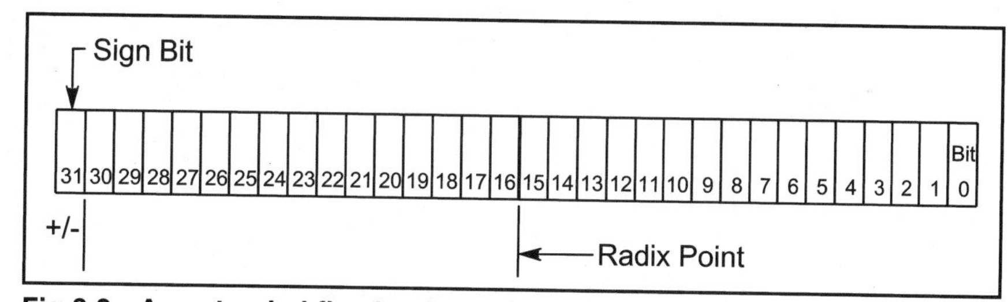

Fig 3.2—An extended fixed-point representation.

$$g \,.\, h = (c + e) \,.\, (d + f) \tag{19}$$

Remember that the fractional parts should be added first so that any carry may be added with the integer parts. Two such numbers may be multiplied in a processor supporting b-bit multiplicands and a 2b-bit product register using:

$$(c.d)\,(e.f) = \left[ce+\mathfrak{I}(cf)+\mathfrak{I}(de)\right] . \left[df+\mathcal{F}(cf)+\mathcal{F}(de)\right] \tag{20}$$

where the script letters indicate the integer and fractional parts of inside products cf and de. Again, the fractional part may overflow; it should be computed first.

Floating-Point Format

Dynamic-range limitations of fixed-point processing often force the use of alternate numeric representations. *Floating-point* format vastly extends the range of numbers that may be represented with a given number of bits, b. In this format, numbers are represented by both a fractional part and another part that stands as a scaling factor. The fractional part, M, is known as the *mantissa*, and the scaling factor, c, is the *characteristic* or exponent. In the most common representation, a positive number F is understood to be:

$$F = 2^C M \tag{21}$$

where the mantissa is restricted to the range:

$$\frac{1}{2} \leq |M| < 1 \tag{22}$$

That restriction means the representation is *normalized*. Note that the characteristic may be either positive or negative, implying a very large range of numbers may be represented. M may be treated as a fixed-point, signed fraction and c as a signed integer.

Two floating-point numbers are multiplied by first multiplying the two mantissas as fixed-point fractions, then adding the characteristics. Since the product of the mantissas will fall in the range:

$$\frac{1}{4} \leq M' < 1 \tag{23}$$

a normalization of the new mantissa and corresponding adjustment of the characteristic may be required. When $M' < 1/2$, it is multiplied by two (shifted one bit to the right) and the characteristic incremented. This check must be performed after every mathematical operation that produces a new result.

Adding two floating-point numbers is done by denormalization of the smaller number. Its bits are shifted left and its characteristic decremented until it is equal to the larger number's. The mantissas are then added directly and the numbers renormalized. From this discussion, it is evident that the mantissa may exceed available register length for both multiplication and addition by a long way; in fixed-point format, that was only the case for multiplication.

Truncation and Rounding

Truncation is the process of "chopping off" one or more LSBs to make a number fit into a smaller register. This may occur, for example, during multiplication of two 16-bit numbers—whose product is a 32-bit number—when only 16 bits are available to store the product. Truncation obviously loses information about the exact value of that product; thus, computational accuracy is degraded. The effect is cumulative over all subsequent operations on that and other numbers and results in quantization noise at the final output. In general, the amplitude of this noise is directly proportional to the number of truncations; that is, the noise is N times worse for an N-multiplication algorithm than for one multiplication.

A reasonably simple way of combating truncation noise is to employ *rounding* of products instead. *Convergent rounding* adds a value of half the kept LSB to the product prior to truncation. In this way, the truncated number is guaranteed to be the value closest to the actual product of those available. Error still exists, obviously, but it is absolutely minimized. Because the error is now just as likely to be positive as negative, and just as likely to be large as small, the resulting quantization noise has a zero mean and is uniformly distributed across the sampling bandwidth.

In the case of truncation, errors do not necessarily have a zero mean and their peak-to-peak range is the same for both positive and negative numbers. Truncation always increases the magnitude of a negative number, though, and always decreases the magnitude of a positive number. Rounding limits the error for each operation to $\pm 1/2$ LSB with a nearly-zero bias, independently of the sign of the number being rounded.

Normalization and Block Floating-Point Notation

Normalization is the process of conditioning numbers to fit a certain scale or reference. As demonstrated above, floating-point numbers go through this process at every stage of their use. *Block floating-point* representations normalize numbers based on their range over a contiguous block or sequence. Block floating-point is a useful notation for many DSP algorithms, especially those that operate on multi-dimensional data arrays.[12]

The characteristic or exponent in block floating-point notation is set equal to that of the largest-magnitude number in a block of, say, N numbers:

$$e = \log_2\left[\max\left(\left|x_k\right|\right)\right],\ 0 \le k \le N-1 \tag{24}$$

where e is an integer. This exponent is associated with a set of signed, fixed-point fractions in binary that are normalized by the factor 2^e. That is, the exponent e is associated with a length-N block of numbers.

An array of length L containing block floating-point vectors (sequences of complex numbers) of length N may be constructed that has L exponents and LN mantissas. This *segmented block floating-point* representation may allow a wider dynamic range than that obtained with a single exponent. It is well-suited to algorithms in which the scale of the data is expected to rapidly change from block to block. Such situations are commonly found in speech processing and discrete Fourier transforms.

Note that with $N = 1$, we just have traditional floating-point representation, wherein normalization must occur at every step. In segmented block floating-point, the normalization occurs on demand and its implementation is subject to change based on knowledge of how data change from block to block. For example, if data magnitude is known to slowly increase, then scale may be adjusted slowly from block to block. This avoids having to recalculate normalizations for each sample, thus saving processing time.

Reduction, Saturation and Justification

Reduction is the normal consequence of binary operations that produce overflow: Just let the sum or product roll over in two's-complement notation, modulo 2^b where b is the number of binary bits used. *Saturation* involves an alternate way to handle overflow: When overflow occurs, retain as the result the maximum binary number possible of the appropriate sign. For 16-bit, fixed-point fractions, the two saturation values are $7FFF and $8000, where the dollar sign indicates hexadecimal notation. Fixed-point DSPs quite often include saturation capability because it is a common tool in speech processing, angular modulation and demodulation, and a variety of other fields. DSP algorithms that employ saturation may tolerate some level of overflow in preceding and following stages. Reduction arithmetic is intolerant of overflow and it must never be allowed to occur.

A computer program may be able to detect that overflow has occurred using *sticky overflow bits* that are latched in a microprocessor until purposefully cleared. A programmer may thus make provision for sensing an overflow catastrophe even though he or she does not necessarily have the information necessary to remedy it by saturation or other means.

In fixed-point notation, numbers are *left-justified*: Any number of zeros may be added at the right-hand side of the number without altering its value. Doing so would extend the *precision* of the number, but not its accuracy since no actual information has been added. On a 16-bit machine, for example, a 16-bit *single-precision* number may have 16 zeros appended on the right to produce a 32-bit *double-precision* number. This conversion applies equally well to all representations. Conversion back to single precision must occur either by truncation or by rounding. As noted above, convergent rounding avoids dc bias terms that may be significant in many applications, but it involves additional computing time.

Unsigned numbers are often just integers and are *right-justified*. Hence, any number of zeros may be appended to the left-hand side without altering the result. Such *extension* does not really extend the precision of a number and it obviously does not affect its accuracy. Zeros appended to a left-justified number are *significant digits*; those appended to a right-justified number are not significant digits. Extension of right-justified numbers, though, may be necessary in DSP algorithms to make them play correctly.

Finding the Logarithm of a Binary Number

Floating-point and block floating-point require the computation of the integer part of the base-two logarithm of a number. This is defined as the scale of the number. A simple algorithm illustrates how this is done in digital computers.

Let us begin with an unsigned, 8-bit binary integer $M = 0010\ 1011_2 = 43_{10}$. As this is a right-justified number, we know zeros appended at the left of M are not significant digits and may be discarded without altering the result. Looking at the string of bits from left to right reveals the position at which the first binary 1 appears. For number M = 43, that is the third digit from the left. The integer part of $\log_2 M$ is therefore $k = \log_2 (0010\ 0000_2) = 8 - 3 = 5$.

If we wish to compute the base-two logarithm of M more accurately, a 256-entry look-up table clearly does the job quite quickly since M is an 8-bit number. The accuracy of a result so obtained is determined by the bit-resolution of the entries, not the number of entries. Were M a 16-bit number, a look-up table of 2^{16} = 65,536 entries would be required. This may tax available memory in embedded systems and another approach is often sought. As shown in the following algorithm, a normalization process allows reduction of table size where facilities exist for fast fractional division.

As an example, take a 16-bit number $M = \$6978 = 0110\ 1001\ 0111\ 1000_2 = 27{,}000_{10}$. Normalize the number to a scale of 8 by dividing it by 2^8 and taking the integer part. We get $N = \text{int}(M / 2^8) = \$69 = 0110\ 1001_2 = 105_{10}$. Now $\log_2 (2^8 N) = 8 + \log_2 N$ is close to the correct answer and $\log_2 N \approx 6.714246$ may be looked up in a 256-entry table. A closer estimate may be computed using the relation $\log_2 M \approx 8 + \log_2 N - \log_2 (2^8 N / M)$. The division is a 16-bit division but only 8 bits of the quotient are retained. Those 8 bits are convergently rounded and again form an address into a 256-entry table; the last term is subtracted to get the final result: $\log_2 M \approx 8 + \log_2 105 - \log_2 (255/256) = 14.7199$. 255/256 is the closest 8-bit fraction to $2^8 N/M$ in this case.

With 16 bits available for the result, the integer part of the result (14) takes up four bits since $14_{10} = 1110_2$; 12 bits remain for the unsigned fractional part. The closest 12-bit fraction is $0.7199 \div 2949/2^{12} = 0.7200$ and the final result (14.7200) comes out a little low. The actual error is about -6.7×10^{-4} or less than 0.01%. The scale of the result is four and in binary, the radix point is placed between the integer and fractional parts: $14.7200_{10} = 1110.1011\ 1000\ 0101_2$.

This example illustrates most of the principles outlined in this chapter. In the next chapter, a significant new set of DSP algorithms is explored: digital filters.

Digital Filtering

The ability to construct very-high-performance filters is a compelling reason to use DSP in radio design. Quite often, expensive analog components may be eliminated in favor of superior DSP implementations.

As filtering requirements get more stringent, filters must get more complex. In the analog world, a filter becomes more complex by adding inductors and capacitors, for example, and the sensitivity of its frequency response to exact element values becomes more critical. Establishing and maintaining exact values over temperature and time quickly become implausible for larger filters. DSP filters, on the other hand, use delay elements and multipliers that may be very accurately set; once they are set, they are unchanging. They are numbers stored in a computer that are not susceptible to temperature or aging problems.

That means filters judged impossible in the analog world may readily be achieved in the digital. Filters having linear phase responses may be constructed, which is a distinct advantage when it comes to digital transmission modes, as described in the following chapter. This is, again, a difficult goal for the analog designer. In addition, cascaded filters may often be numerically combined into a single filter, saving computation time—that quantity that is always obtained at a premium.

DSP filters come in several varieties. Each has an analog counterpart that, although incredibly hard to build, would function nearly the same. That fact may be utilized to advantage in understanding how they function. The approach below even shows how traditional analog filter families may be adapted to digital use; however, certain DSP filter constructs are not normally attempted in analog.

Characterization of Signals: Terminology of Linear, Time-Invariant Systems

A system or function $y_t = H(x_t)$ is defined as linear if and only if it is *commutative*. That is:

$$H\left(A_1 x_{1_t} + A_2 x_{2_t}\right) = A_1 H\left(x_{1_t}\right) + A_2 H\left(x_{2_t}\right) \tag{25}$$

where x_1 and x_2 are samples taken at different times; A_1 and A_2 are their amplitudes. A system is *time-invariant* if and only if older output samples are determined only by older input samples. The following equation must then be true:

$$Y_{t-t_0} = H\left(x_{t-t_0}\right) \tag{26}$$

H is often called the *system* or *transfer function.*

A system may be defined as *causal* if its output depends only on current and past input values. A system, for example, whose past outputs changed based on new inputs would not be causal. A system is stable if and only if a bounded set of input values produces a bounded set of output values. This is the same as saying the *impulse response* of the system integrates to a finite value.

Impulse Response

All filters may be characterized by their impulse responses. The impulse response of a filter is its output when the input is a one-sample, unity-amplitude impulse; think of this input as a very narrow "spike." Output may be quite complex, as in **Fig 4.1**; this is often referred to as "ringing," although it is just a consequence of how filters—both analog and digital—behave. Output voltage vs time may be sampled, just as any other analog signal may be; however, the sampled impulse response cannot be infinite in length. It is, therefore, only an approximation to that required to exactly describe the filter. The truncated length of the sampled impulse response is found to correspond with an error in the digital filter's frequency response—the thing in which main interest lies. A digital filter designed this way is therefore called a *finite-impulse-response (FIR)* filter.

Imagine building an analog low-pass filter and shooting a unity-amplitude impulse into it, making the width of the spike very narrow. Then, take L samples of the output waveform at regular intervals $1/f_s$. The sampled impulse response may be used as a sequence of *coefficients* in a basic FIR filter structure employing delay elements and multipliers, as shown in **Fig 4.2**. Each box labeled z^{-1} is a one-sample delay; the cascaded string of those boxes is an *(L – 1)-sample delay line*. Programmers will recognize that this is just a buffer of length L – 1. Each location in the delay line may be referred to as a *tap* in the line. The

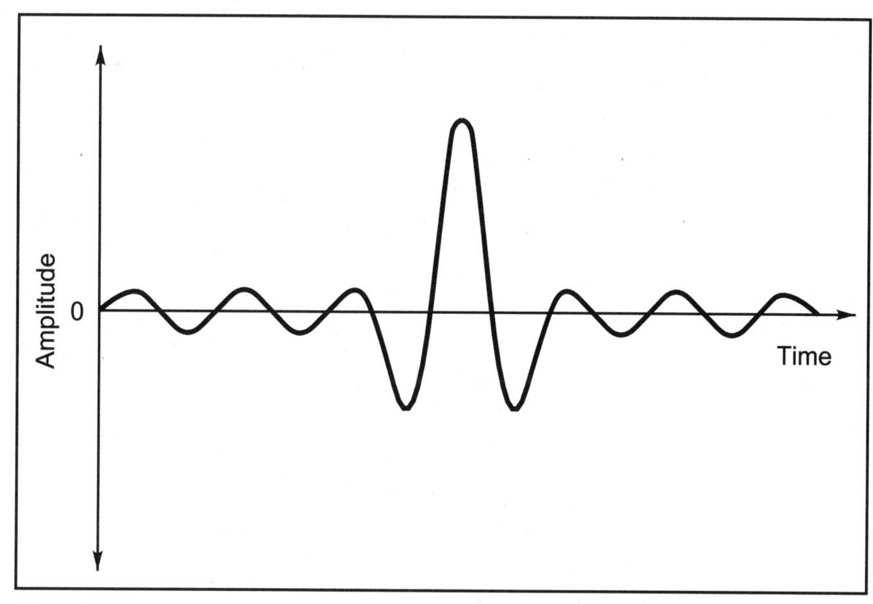

Fig 4.1—Impulse response of a typical filter.

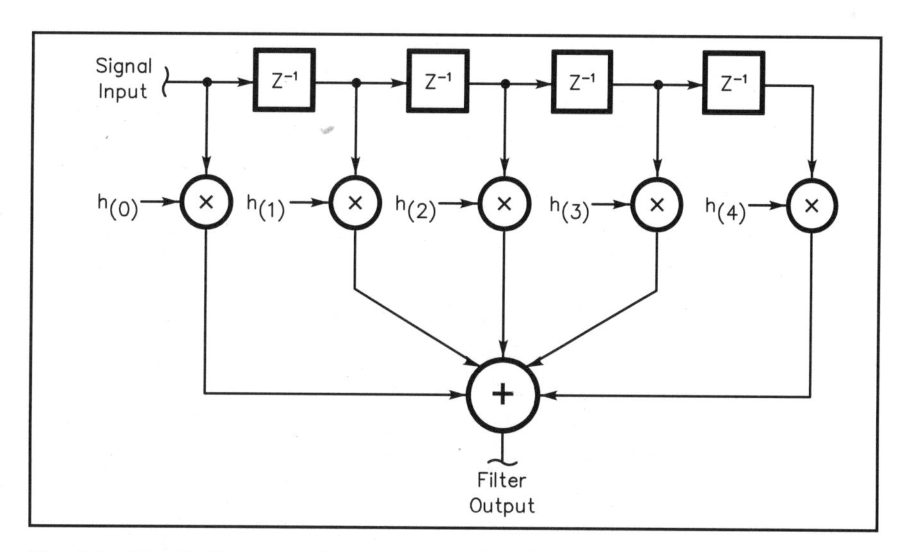

Fig 4.2—Block diagram of a short FIR filter.

datum at each tap, x_n, is multiplied at each sample time with one of the coefficients, h_n. All the products are summed at each sample time to produce the filter's output.

At the next sample time, samples are right-shifted down the delay line by one position and the *multiply-and-accumulate (MAC)* operation is performed again. Coefficients remain in place and do not shift. The mathematical expression describing this repetitive operation is also called a *convolution sum*:

$$y_k = \sum_{n=0}^{L-1} h_n x_{k-n} \tag{27}$$

where y_k is the output at sample time k, x_{k-n} is the set of input samples, and h_n is the set of L coefficients. Since the output depends only on past input values, the filter is a causal process. Since no feedback is employed, it is unconditionally stable. The sampled input signal is *convolved* with the filter's impulse response; the output spectrum is the product of the two input spectra. This relationship was illustrated in Chapter 2: Convolution in the time domain corresponds to multiplication in the frequency domain. Inversely, multiplication in the time domain (mixing) corresponds to convolution in the frequency domain.

Computer Design of FIR Filters

In any filter-design project, a desired frequency response is usually sought and element (coefficient) values must be computed. Most methods begin with an estimate of the number of elements needed to achieve the desired response. In the case of FIR filters, Rabiner and Gold[13] indicate the number of taps, L, must be at least:

$$L \geq \frac{10\log(\delta_1\delta_2) - 15}{14\left(\frac{f_T}{f_s}\right)} \tag{28}$$

where δ_1 is the passband ripple, δ_2 is the stopband ripple, f_T is the transition bandwidth, and f_s is the sampling frequency. This equation assumes enough bits of resolution are used to achieve the required accuracy. It is shown below that truncation of filter coefficients affects frequency response adversely and unexpected things may occur.

Normally, an FIR filter's impulse response has a symmetry about center, such that $h_0 = h_{L-1}$, $h_1 = h_{L-2}$, and so forth. This is sufficient to ensure a linear phase response and flat group-delay characteristics. The total delay through an FIR filter of length L is:

$$t = \frac{L}{2f_s} \tag{29}$$

This delay is *independent* of input frequency: That is why the filter has a linear phase response.

One FIR filter design approach takes advantage of the fact that a filter's frequency response is just the *Fourier transform* of its impulse response. Fourier transforms are discussed in Chapter 8. Filters may be designed starting with a sampled version of the desired frequency response and an *inverse Fourier transform* employed to obtain the impulse response. Better designs may be produced in many cases using an algorithm developed by Parks and McClellan.[14] It achieves an *equi-ripple* design in which all the passband ripples are the same amplitude, as are all the stopband ripples. Another popular algorithm is called the *least-squares method.* Its claim to fame is that it minimizes error in the desired frequency response.

Since finding coefficient sets for a given filter design is so computationally intensive, it is a good job for a computer. DSP filter-design programs are readily available at low cost. Some of those are mentioned in Chapter 11 and in the Bibliography.

Infinite-Impulse-Response (IIR) Filters

While FIR filters have a lot going for them, they tend to require a large number of taps for decent transition bandwidths and an attendant amount of processing power. As opposed to that, an *IIR* filter may provide sharp skirts with relatively few calculations. What it will not provide, in general, is a linear phase response. In circumstances where the computational burden is of more concern than the phase response, IIR filters may be desirable.

Unlike FIR filters, IIR filters employ feedback: That is what makes their impulse responses infinite. The same thing is true of traditional analog filter types, such as Chebychev and elliptical. For that reason, IIR filters are usually designed by converting analog prototypes. IIR filters may have both zeros and poles; FIR filters have only zeros.

The transfer function of an analog Chebychev low-pass filter may be written as the ratio of a constant to an nth-order polynomial:

$$H_s = \frac{K}{a_0 s^n + a_1 s^{n-1} + a_2 s^{n-2} + \cdots + a_n s^0} \tag{30}$$

Tables in the literature, such as Zverev[15], list values of coefficients a_n related to the cutoff frequency; these may be translated into component values in the actual filter. This low-pass design may be transformed into band-pass or band-stop responses.

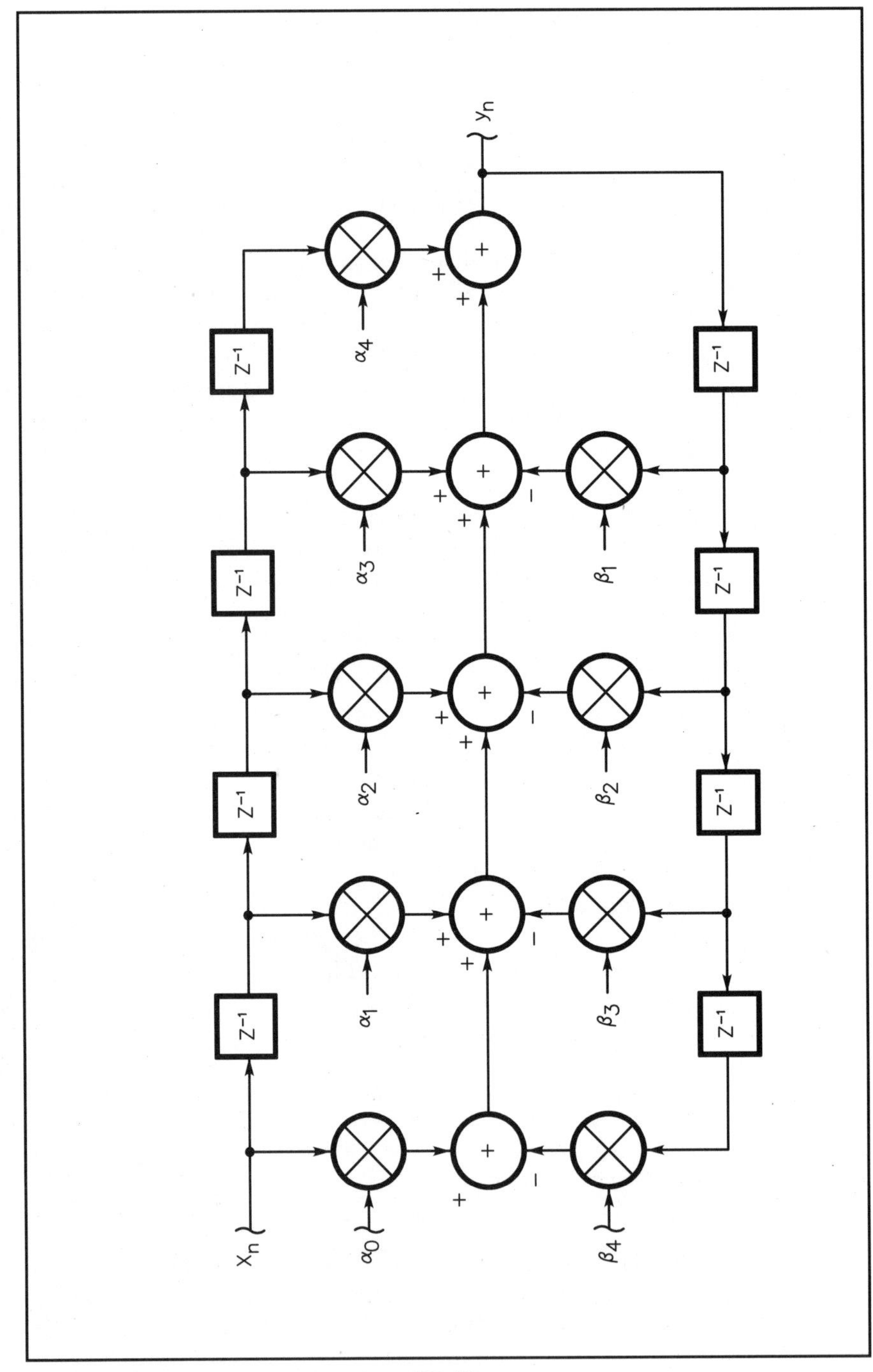

Fig 4.3—Block diagram of a short IIR filter.

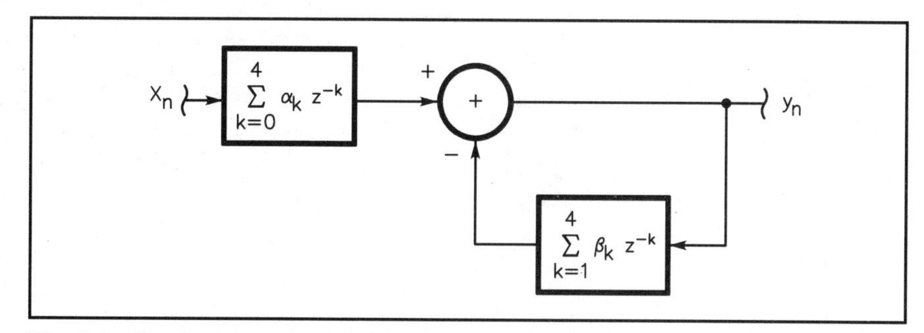

Fig 4.4—Equivalent block diagram of a cascade-form IIR filter.

Two popular methods exist for deriving the digital transfer function from the analog: These are known as the *impulse-invariant* and *bilinear transform* methods.

The impulse-invariant method assures that a digital filter will have an impulse response equivalent to its analog counterpart, and thus, the same phase response. Problems arise, though, if the bands of interest are near half the sampling frequency. The digital filter's response may develop serious errors in this case. While most filter-design software is capable of this method, it is not as often used as the bilinear-transform method.

The bilinear transform makes a convenient substitution for s in Eq 30 and the filter comes out looking like:

$$Y_k = \sum_{n=0}^{L-1} \alpha_n x_{k-n} - \sum_{n=1}^{L-1} \beta_n y_{k-n} \tag{31}$$

This filter has L zeros and L – 1 poles. The block diagram of such a filter for L = 5 is shown in **Fig 4.3**. Feedback is evident in both the equation and diagram since paths involving coefficients β loop back and are added to the signal path. See **Fig 4.4**. The *direct form* of Eq 31 may be factored into 2-pole section and implemented in cascaded form. This configuration requires a few more multiplications than the direct form, but it suffers less from instability problems that plague IIR filters. Since feedback is being used, IIR filters are not necessarily unconditionally stable. They are prone to *limit cycles*, or low-level oscillations sustained by quantization effects described in the previous chapters.

Numerical Effects in Digital Filters: Coefficient Accuracy

When computers are used to design DSP filters, coefficients are usually represented in floating-point format to the full accuracy of the computer, often with 12 or more significant decimal figures in the mantissa. Embedded, fixed-point implementations ordinarily achieve only 16-bit binary accuracy. The truncation or rounding of coefficients and data to this resolution affects the frequency response, ultimate attenuation and noise performance of digital filters.

It is interesting to note that while coefficient accuracy affects frequency response, it does not contribute to quantization noise in a filter's output signal since the noise sources are not processed at all by the system. On the other hand, truncation and rounding of data do not affect frequency response but add noise to the output. Notice that the product of two 16-bit numbers is a 32-bit number and many of these must be added together to form the output of an FIR filter. The result may grow by several more bits before a final result is produced. At some point, the result may overflow the final accumulator, especially in FIR filters with low shape factors. When the input is a strange, sign-matched copy of the filter's impulse response, worst-case output may grow as large as the sum of the absolute value of all the coefficients:

$$y_{max} = \pm\sum_{n=0}^{L-1}|h_n| \tag{32}$$

Data, coefficients, or both might have to be scaled by the reciprocal of this number to avoid overflow. Usually, the final output value must be truncated or rounded to some number of bits, say 16. That introduces a small additional quantization-noise component that has already been defined.

Some systems may not have the luxury of a final accumulator that has 32 or more bits. In this case, individual products in an FIR filter must be rounded prior to accumulation. To analyze data quantization noise in such an FIR filter, a modified block diagram is used that inserts noise sources e_{n_1}, e_{n_2} and so forth at the point where individual products are rounded. See **Fig 4.5**. An input scaling factor F may also be added to prevent overflow. Clearly, each noise source adds directly to the output. Assuming the noise sources are not correlated to one another, the total noise output is:

$$e_{total} = \sum_{k=0}^{L-1} e_{kn} \tag{33}$$

After the quantization-noise derivation in Chapter 2, the variance of the output noise for rounded products is equal to the normalized noise power:

$$\sigma_e^2 = L\left(\frac{2^{-2b}}{12}\right) \tag{34}$$

where b is the number of bits to which interim results are rounded.

The effects of *coefficient quantization error* are more difficult to analyze

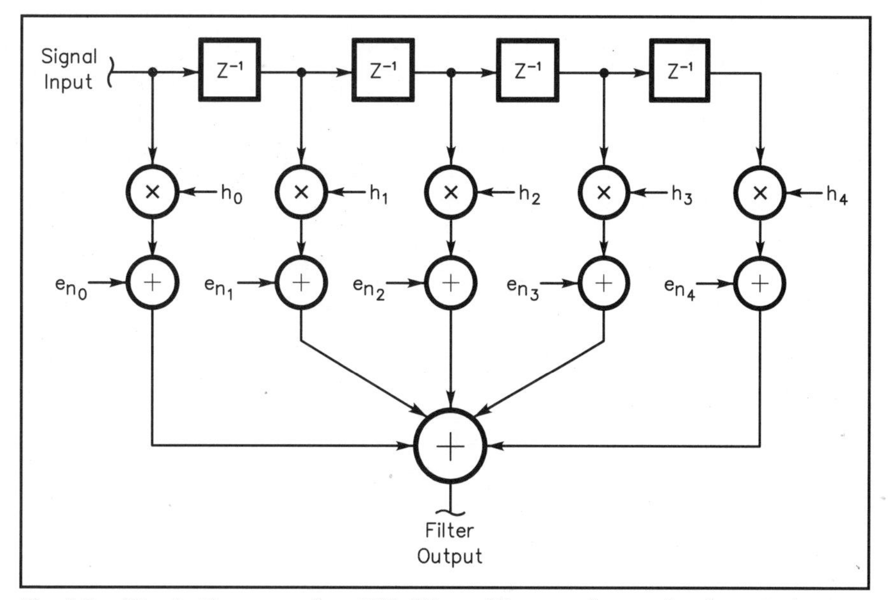

Fig 4.5—Block diagram of an FIR filter with rounding noise inserted.

mathematically, but the model is still fairly easy to draw. Refer to **Fig 4.6A**. Here, the error sources reside in the coefficients themselves; however, the errors are constants and do not change from sample to sample as data quantization errors do. The coefficients never change, so neither do the errors. Each interim product may be separated into the product involving the coefficient and the product involving only the error. See Fig 4.6B. A final refinement to this model shows that this produces a small bias in frequency response because the system is the same as an FIR filter with infinite coefficient accuracy in parallel with one that uses only the errors as its coefficients, as in Fig 4.6C. The perfect FIR filter has the desired frequency response, while the error filter has an undefined response. This shows that coefficient truncation or rounding introduces distortion in a filter's frequency response, but not noise in its output. It also demonstrates that filters designed in floating-point format but for use in fixed-point systems must be checked with actual coefficient resolution. Further, it is shown that when errors in coefficients can be determined to be repeatable, tactics may be employed that minimize them. That is the subject of ongoing research.

IIR Limit Cycles

Limit cycles in IIR filters were mentioned above as a quantization prob-

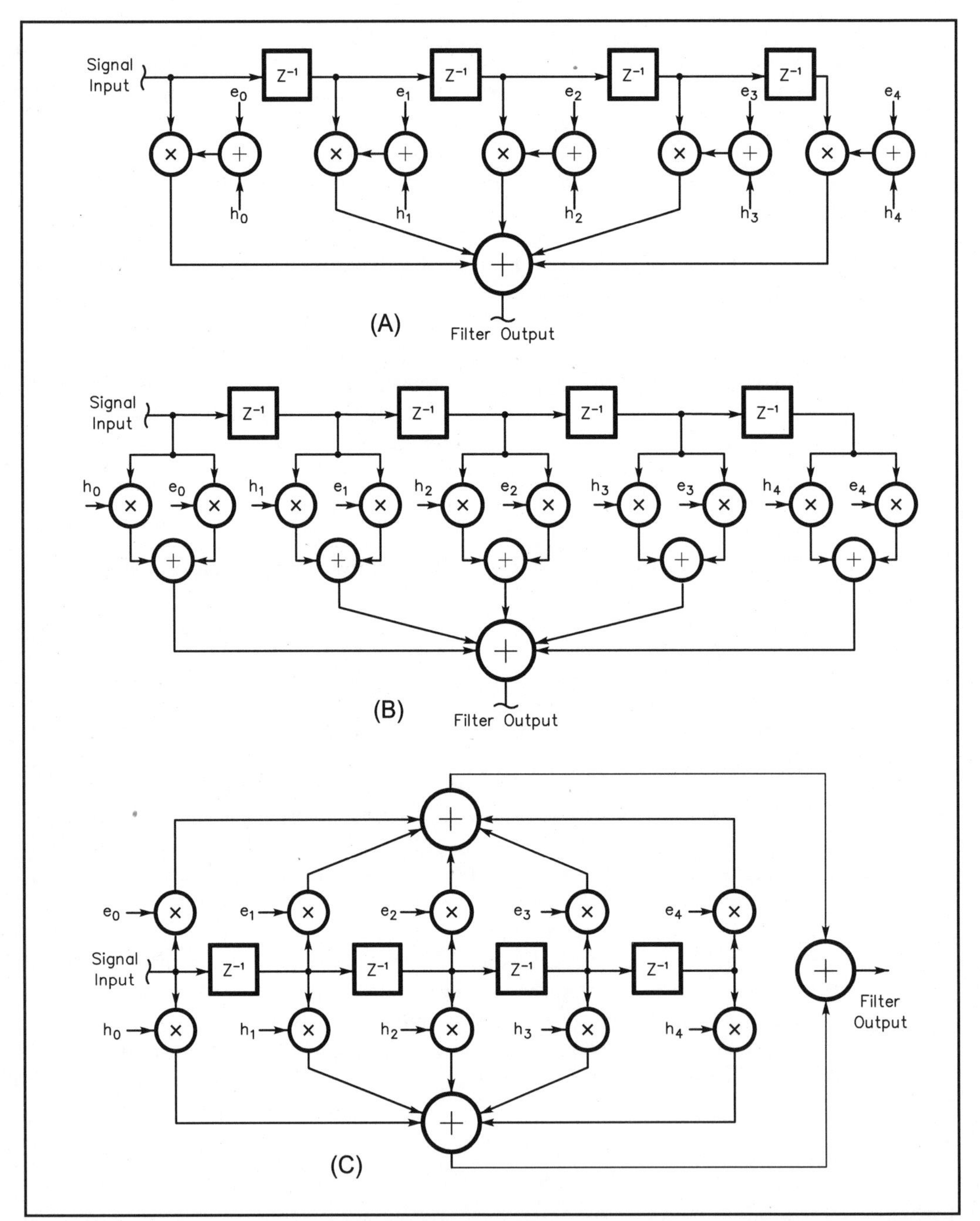

Fig 4.6—At A, block diagram of an FIR filter modeling coefficient errors. At B, modified block diagram showing separation of error products. At C, final block diagram showing equivalence to two separate filters whose outputs are summed.

lem. The presence of feedback in the algorithm poses this problem at signal levels near the smallest-resolvable signal. Suppose the algorithm is started with zero at its input and with a very small number at a feedback node. Were the coefficients sizable enough, and depending on the complexity of the filter, a very small numerical error might propagate through the system endlessly because it never made it to zero in any multiplication or addition. This cannot happen in a straight FIR filter because signals do not find their ways back to the input. *Adaptive FIR filters* are an exception, covered in Chapter 8. Limit cycles also will not happen in an IIR filter when coefficient values are sufficiently low to assure that two small numbers, when multiplied, produce a zero product.

Limit cycles exhibit a dead band and other familiar characteristics of oscillators, but in the realm of non-linear, step-wise behavior only. Further details will not be discussed here, except to point out that with due care, these oscillations need not exceed several LSBs. Now the reason for wanting to factor IIR filters into cascaded, 2-pole sections becomes evident.

Floating-Point Effects

Floating-point format readily removes dynamic-range limitations of fixed-point, but it also suffers from effects of finite precision. In computing interim products in an FIR filter, each product must be renormalized and may lose precision in the process. The block diagram of a model for this, **Fig 4.7**, is quite similar to Fig 4.6A; they differ in that errors in products are multiplicative and not additive. Distortions are therefore both those of the output signal and of frequency response.

Derivation of an expression for these floating-point errors is a compli-

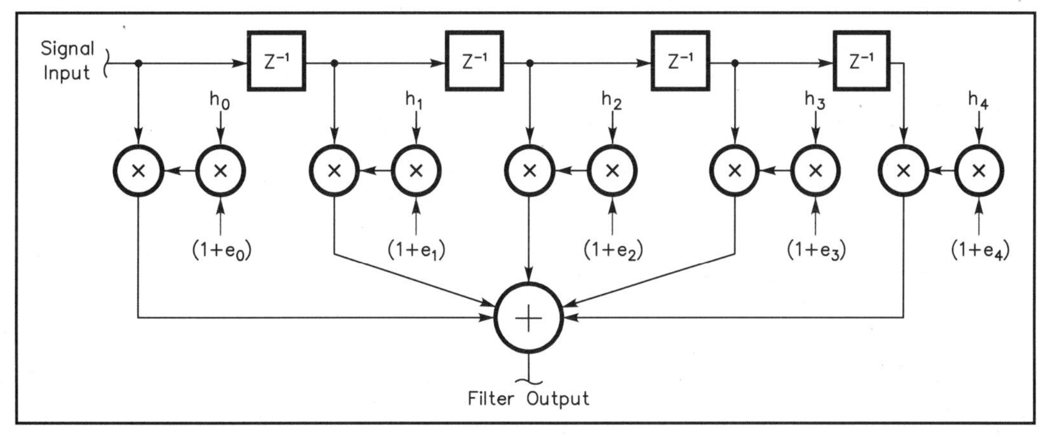

Fig 4.7—A model for errors in floating-point implementation of DSP filters.

cated session in statistics. Fortunately, Oppenheim and Schafer[16] have done it for us and showed that output SNR is bounded by:

$$\frac{P_{signal}}{P_{noise}} \leq \frac{(3)\left(2^{2b}\right)}{(L+1)} \tag{35}$$

Filter Design Using Impulse Windows

An *impulse window* is just a sequence whose envelope matches some particular shape, such as a rectangle or a triangle. Rectangular and triangular windows are shown in **Fig 4.8** for L = 64, along with several other shapes and their frequency responses when used as coefficients in an FIR filter. It is evident that the rectangular window achieves the fastest roll-off to stopband and they all produce various amounts of ultimate attenuation and stopband ripple.

Note that the positions of zeros in the frequency responses are dependent on the length of the window and its shape. These functions may be selected according to the demands of a specific application. These low-pass functions may be transformed to band-pass and high-pass responses as shown below. Band-pass follows directly from the low-pass case. Transformation takes place through multiplication (mixing) of the low-pass prototype's impulse response by a sinusoidal, "local-oscillator" sequence. This is precisely the same multiplication and mixing that takes place in an analog mixer; but now, we are concerned with the frequency response of a filter, not the frequency of a signal. The general frequency-translation properties of multipliers and filters are treated specifically in the following chapter.

Let us say that the prototype LPF has coefficients h_n and a frequency response H_ω. Multiplying the coefficients by a sinusoid ω_0 results in new coefficients:

$$h_n' = h_n \cos\left(\omega_0 n t_s\right) \tag{36}$$

As will be proved later, the frequency response of this filter is:

$$H_\omega = \frac{H_{(\omega-\omega_0)} + H_{(\omega+\omega_0)}}{2} \tag{37}$$

which is a band-pass filter centered at ω_0. To perform this transformation on the L coefficients of the prototype, calculate new coefficients according to:

$$h_n' = h_n \cos\left[\omega_0 \left(n - \frac{L}{2} + \frac{1}{2}\right) t_s\right] \tag{38}$$

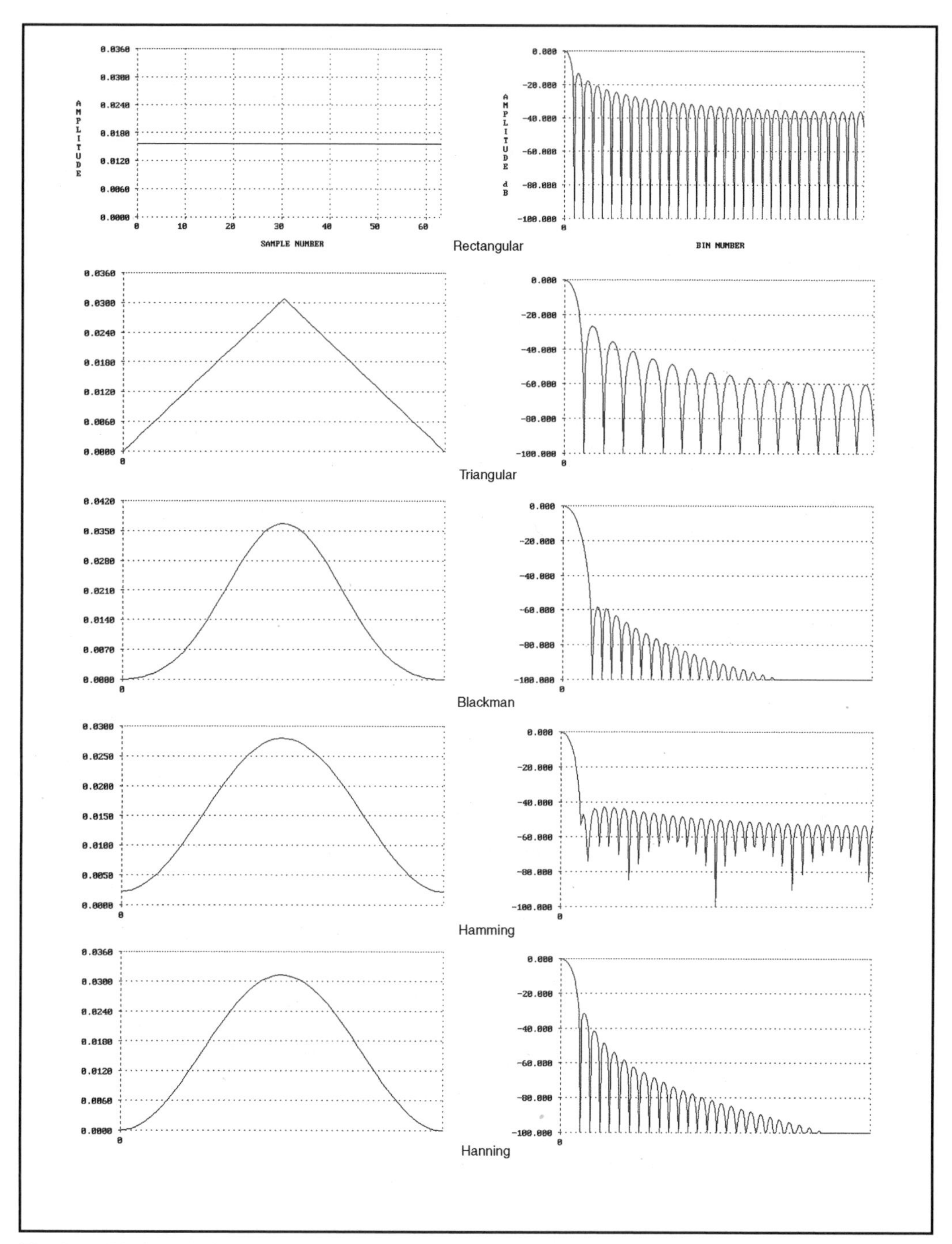

Fig 4.8—Various window functions and their Fourier transforms.

Fig 4.9—A notch filter formed using a BPF and simple subtraction.

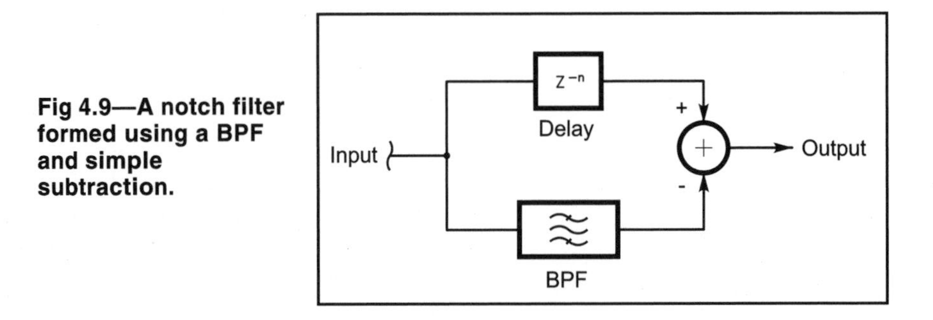

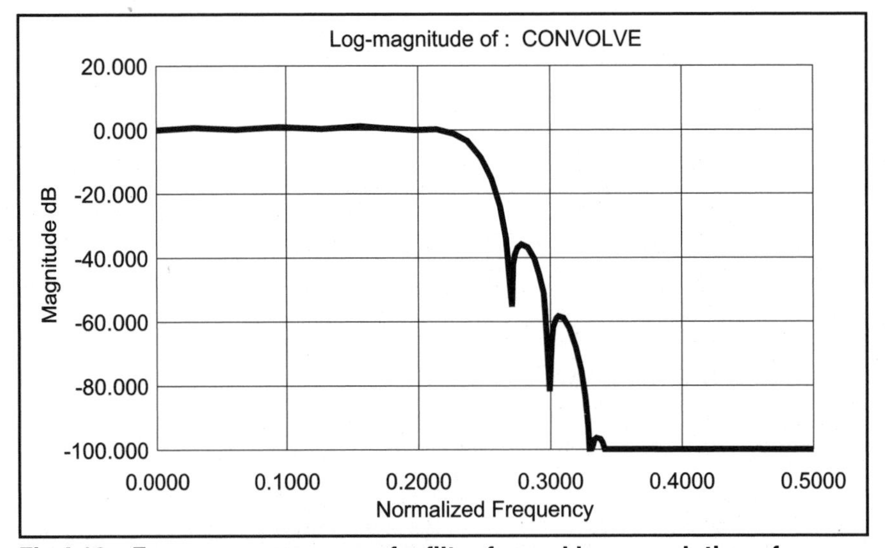

Fig 4.10—Frequency response of a filter formed by convolution of rectangular and Blackman windows.

In going to high-pass, the frequency response of the prototype LPF is translated by half the sampling frequency, or $f_s/2$. Now the local oscillator can only be a square wave of amplitude ± 1, since it is at the Nyquist frequency. In the multiplication, this is equivalent to multiplying every other coefficient by –1. This is perhaps the simplest transformation possible on the prototype LPF.

Band-stop response may be obtained from the band-pass case above by simply subtracting the band-pass filter's output from the broadband input. Compensation must be made for the delay through the filter, though, so a delay line of $L/2f_s$ samples is included in the broadband signal's path. See **Fig 4.9**. It is evident from this diagram that multiple notches or passbands are readily supported by combining the outputs of several different BPFs.

Filters may even be convolved with each other to produce hybrid shapes and responses. For example, a rectangular window may be convolved with a Blackman window to get the benefits of both: the fast roll-off of the rectangular and the good ultimate attenuation of the Blackman. The frequency response of such a filter is shown in **Fig 4.10**. A rectangular and a Blackman window, each of L = 64, were convolved to produce a new filter of L = 127.

The block diagram of a convolver is the same as that of an FIR filter. Two impulse responses are convolved according to the same equation presented for FIRs, except the arguments are the impulse responses of the filters, g_n and h_n. The two sequences need not have the same length. When g_n has length L, and h_n length M, the new sequence is:

$$h'_n = \sum_{k=0}^{L-1} g_k h_{n-k} \tag{39}$$

For M values of n, it is evident that the new sequence has non-zero length L + M – 1.

Analytic Signals and Modulation

As demonstrated, DSP implementations of complex functions compel designers to reexamine the mathematics behind them. Computers are good at crunching numbers, but they do exactly what they are told. If we wish to modulate or demodulate a signal, therefore, we need to formulate the modulation mathematically and reduce it to discrete processing.

Learning about various modulation formats and their representation by complex numbers began in earnest after James Maxwell published his treatise on electromagnetism (1873). His equations are remarkable in that they either directly or indirectly predict virtually every facet of radio wave propagation. Understanding them, though, need not be so complex. Take a look at Reference 1 for a brief treatment of the math if you need better understanding. A brief review of complex-number algebra is also included there, a working knowledge of which is helpful in the following discussion. That digression to Maxwell points out that mathematics are a great tool when handled deftly.

Among the first modulation formats were on-off keying, a form of AM—although those early transmitters and receivers certainly had some FM, too! Linear AM and eventually FM and PM also took their places on the stage in time. All formats have precise mathematical descriptions that are examined below. We owe our insight into them to many great minds working on chalkboards, then in the field, for a long time. With the arrival of sufficient processing horsepower, DSP has exploited complex-number modulation theory brilliantly by taking advantage of *analytic representations* of signals.

An analytic signal is one that has both a real and imaginary part. It may be an actual, physical signal, such as that from a microphone, that may be sampled at regular time intervals. An audio signal may be the input to a modulator, a squelch circuit, or a compressor. By looking at that signal in the complex plane, the bases for mixing, modulation and other fundamental aspects of theory are explained.

A real signal, such as a cosine wave, may be thought of as having both real and imaginary parts. While that says little about the actual nature of a cosine wave, whose shape is familiar to most, such analytic representation leads to a great many of the formulations that follow. We shall discuss cosine waves of angular frequency $\omega = 2\pi f$. In sampled form, we may define the sequence:

$$x_t = \cos \omega t \tag{40}$$

where t is time. Note that throughout this book, arguments of trigonometric functions such as the one in Eq 40 assume parentheses around all terms prior to and including t. That is, Eq 40 could more clearly be written as $\cos(\omega t)$; but to avoid clutter, they are eliminated here.

In the complex domain, this cosine wave is equal to the sum of two complex signals:

$$\cos \omega t = \frac{1}{2}\left[\left(\cos \omega t + j \sin \omega t\right) + \left(\cos \omega t - j \sin \omega t\right)\right] \tag{41}$$

In the complex plane, where the axes are the real and the imaginary, this signal may be represented by two vectors rotating in opposite directions, as shown in **Fig 5.1**. Across the range of positive and negative frequencies, the cosine wave's spectrum is shown in **Fig 5.2**. This signal has both positive and negative frequency components. Pretty obviously, one term inside the brackets is a positive frequency and the other, negative. Imaginary terms—the *j*sin terms—cancel to make the equation true.

Now this depiction is elegant, but what does it really mean? Well, it means that signals represented in analytic form may be made to have a one-sided spectrum; that is, one having only positive or only negative frequencies. This fact is crucial in understanding the mixing phenomenon.

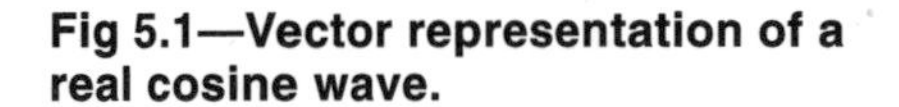
Fig 5.1—Vector representation of a real cosine wave.

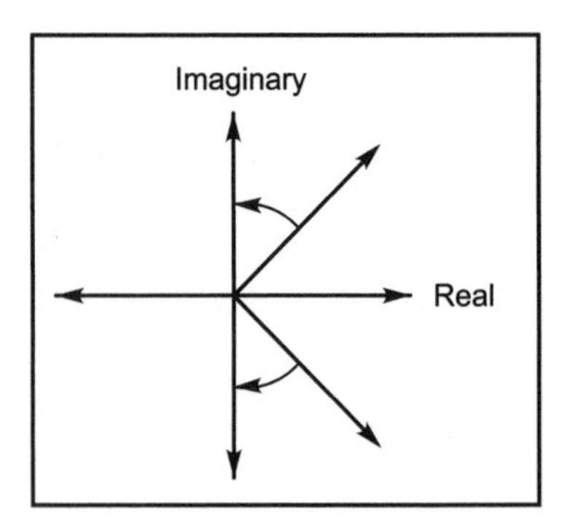

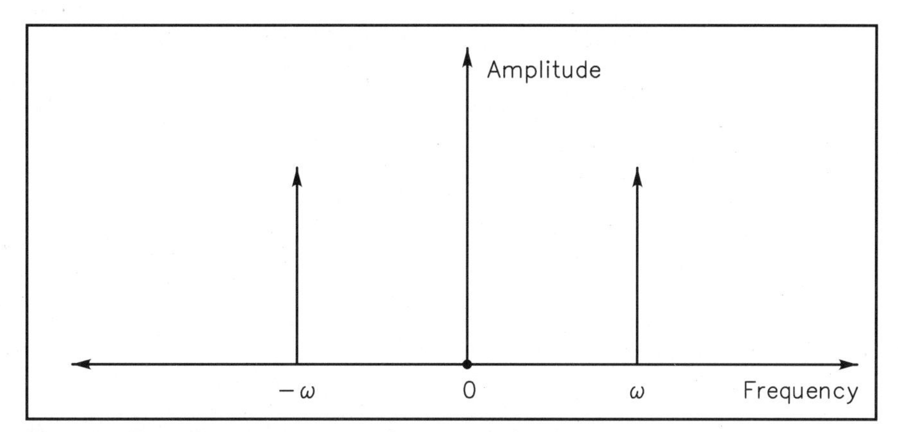

Fig 5.2—Two-sided spectrum of a real cosine wave.

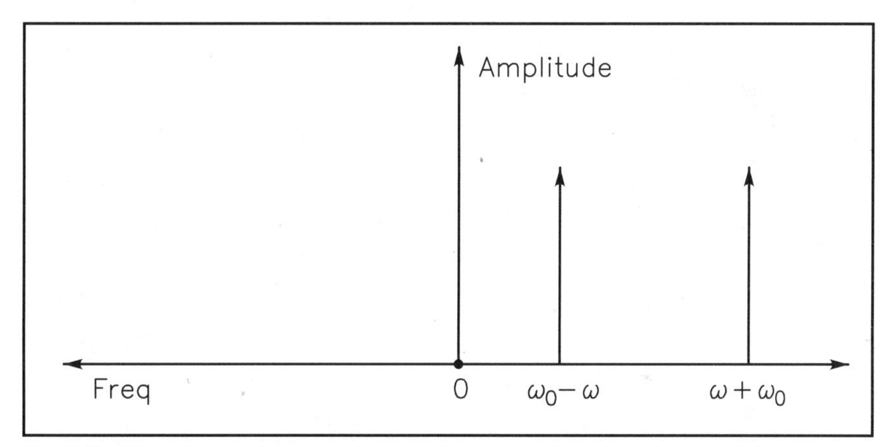

Fig 5.3—Spectrum of a complex cosine wave that has been mixed upward in frequency.

Frequency Translation: Complex Mixing

Were we able to translate the two-sided spectrum of a cosine wave upward in frequency far enough, we would have two positive frequencies separated by twice the original signal's frequency. See **Fig 5.3**. For a real signal, that is exactly what happens in a conventional, analog mixer. Both sum and difference frequencies are generated.

To see this mathematically, we first invoke the identity discovered by Leonhard Euler (1747, history gets it right):

$$e^{j\omega t} = \cos\omega t + j\sin\omega t \qquad (42)$$

This is a more compact notation for complex signals. Our real cosine wave now takes on the form:

$$\cos\omega t = \frac{1}{2}\left(e^{j\omega t} + e^{-j\omega t}\right) \tag{43}$$

When we mix this with a real carrier, say $\cos\omega_0 t$, we get the product of the two inputs:

$$\begin{aligned}
(\cos\omega_0 t)(\cos\omega t) &= \left[\frac{\left(e^{j\omega_0 t} + e^{-j\omega_0 t}\right)}{2}\right]\left[\frac{\left(e^{j\omega t} + e^{-j\omega t}\right)}{2}\right] \\
&= \frac{\left[e^{j(\omega_0+\omega)t} + e^{-j(\omega_0+\omega)t}\right] + \left[e^{j(\omega_0-\omega)t} + e^{-j(\omega_0-\omega)t}\right]}{4} \\
&= \frac{1}{2}\left[\cos(\omega_0+\omega)t + \cos(\omega_0-\omega)t\right]
\end{aligned} \tag{44}$$

Multiplication of two cosine waves produces the convolution of their two-sided spectra. It is interesting to note what Eq 44 says about real, analog mixers and their performance. At an early stage (in the 1930s), engineers realized that *commutating mixers*—ones having switches as their active elements—still behaved according to Eq 44; only in that case, the carrier (local oscillator) is a square wave. A square wave contains a fundamental component and many harmonics. Fourier analysis (Chapter 8) tells us that a real square wave is composed of odd-order sinusoids of the form:

$$y_t = \frac{4\cos\omega_0 t}{\pi} + \frac{4\cos 3\omega_0 t}{3\pi} + \frac{4\cos 5\omega_0 t}{5\pi} \cdots \tag{45}$$

To find out what happens when using this as a local oscillator (LO), we substitute it into Eq 44.

Taking the first two terms of Eq 45 as our LO (the fundamental and third harmonic only), we have mixer output:

$$\begin{aligned}
y_t &= \left(\frac{4\cos\omega_0 t}{\pi} + \frac{4\cos 3\omega_0 t}{3\pi}\right)(\cos\omega t) \\
&= \frac{2}{\pi}\left[\cos(\omega_0+\omega)t + \cos(\omega_0-\omega)t\right] + \frac{2}{3\pi}\left[\cos(3\omega_0+\omega)t + \cos(3\omega_0-\omega)t\right]
\end{aligned} \tag{46}$$

Now the left-hand term represents the sum and difference frequencies of interest and the amplitude of each component is $2 / \pi$. The *conversion loss* of this mixer is therefore:

$$20\log\left(\frac{2}{\pi}\right) \approx -3.9 \text{ dB} \tag{47}$$

For a proof of the spectral content of a square wave using the *mean-squares method* of harmonic analysis, see Appendix A.

SSB Modulation

Let us get back to the case of microphone audio serving as the input to a modulator. Call this sampled signal I_t. Suppose we wish to build an SSB transmitter. We know the audio is a real signal with a two-sided spectrum. It is possible to convert this real signal to an analytic signal with only positive-frequency components—by generating a quadrature signal, Q_t, in which all frequencies are phase-shifted by 90° from those in signal I_t. The signals I and Q are treated as an *analytic pair.* Signal $I_t + jQ_t$ contains only positive frequencies. Negative frequencies cancel each other, while positive frequencies reinforce. A function that phase-shifts all the frequency components by 90° is called a *Hilbert transformer*. Such a thing is very difficult in the analog world, but it is easy in DSP.

We now multiply (mix) this analytic signal with a complex oscillator:

$$y_t = e^{j\omega_0 t} \tag{48}$$

having both cosine and sine parts, and the translated signal takes the form:

$$\begin{aligned} e^{j\omega_0 t}\left(I_t + jQ_t\right) &= \left(\cos\omega_0 t + j\sin\omega_0 t\right)\left(I_t + jQ_t\right) \\ &= \left(I_t\cos\omega_0 t - Q_t\sin\omega_0 t\right) + j\left(I_t\sin\omega_0 t + Q_t\cos\omega_0 t\right) \end{aligned} \tag{49}$$

Since we are interested only in transmitting a real signal, we have to compute only the real part of Eq 49. Ignoring signal delays, such a "half-complex" mixer is implemented as shown in **Fig 5.4**. It produces a USB signal. Were the real and complex parts added instead of subtracted in the final summation, an LSB signal would be the result. An SSB signal ought to occupy only 3-4 kHz, so the next step is to add a band-pass filter that attenuates both low and high audio frequencies prior to the Hilbert transformer and the mixers. A Hilbert transformer may be implemented as an FIR filter structure, with its attendant delay; the design needs a delay in the non-transformed signal path, I. Now the modulator resembles that shown in **Fig 5.5**.

As alluded to above, a Hilbert transformer may be embodied by creating an *analytic filter pair* that combines the functions of filtering and phase shifting. As the design grows, this arrangement will minimize the computational burden in this part of the circuit. Substituting the pair in the block diagram and adding the ADC, DAC and analog anti-aliasing filters that are naturally needed produces the diagram of **Fig 5.6.**

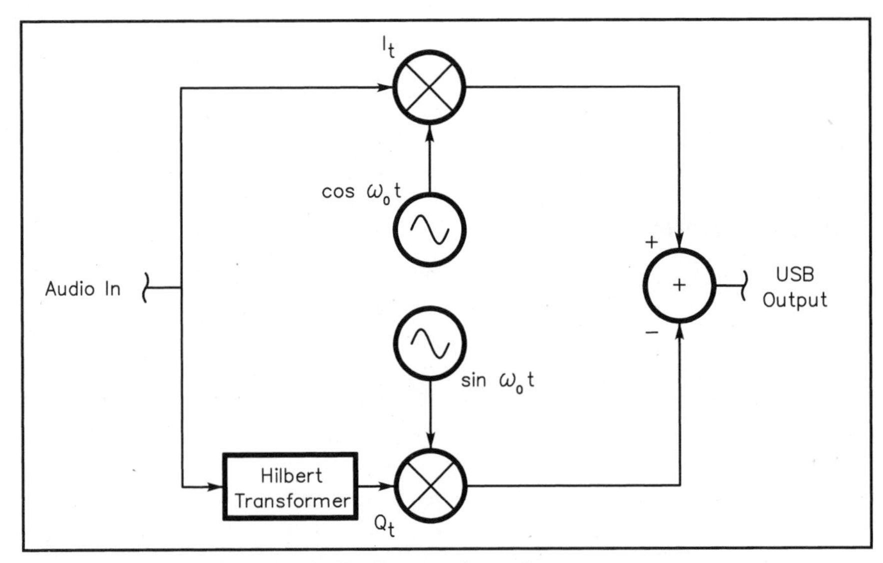

Fig 5.4—Block diagram of a half-complex mixer.

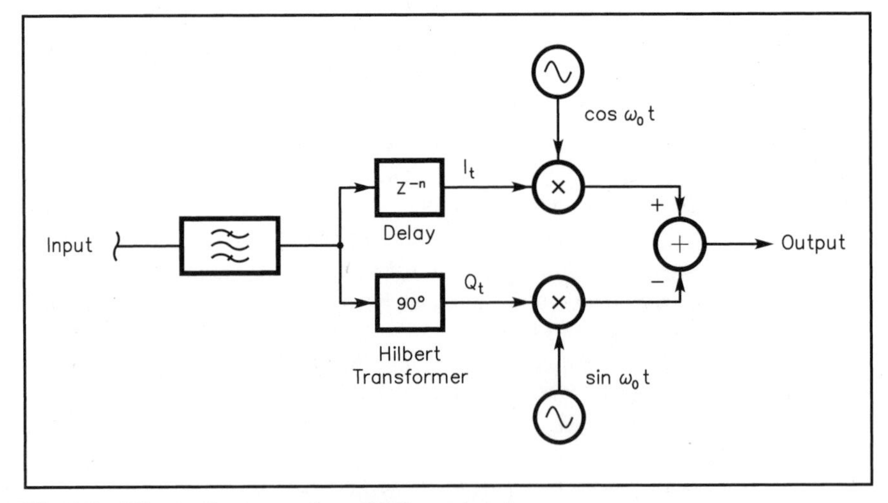

Fig 5.5—Block diagram of an SSB modulator.

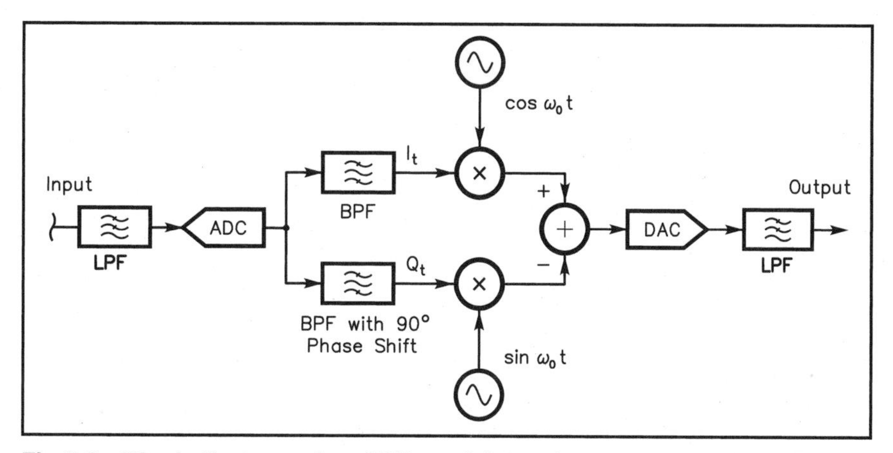

Fig 5.6—Block diagram of an SSB modulator that uses an analytic filter pair.

In most applications, the output (IF) frequency range will be significantly higher than the input range and an increase in sampling frequency is called for, even though the mixer does not alter the limited bandwidth of its input. At the DAC, the sampling frequency must be at least twice the highest frequency component in the output. It is particularly convenient to choose an output sampling frequency that is four times the local-oscillator frequency. The cosine oscillator takes on values of only 1, 0, –1, and 0, thus avoiding actual multiplication in its mixer. This saves time and accuracy.[17] Similarly, the sine oscillator takes on values of only 0, 1, 0, and –1. Complex mixing and computation of the real part of the result is thus made devilishly simple: Output is an unaltered or negated copy of the I signal at even-numbered sample times, and an unaltered or negated copy of the Q signal at odd-numbered sample times.

A sampling-frequency increase requires interpolation filters, as discussed above. Their addition to the block diagram results in the configuration of **Fig 5.7**: a complete digital SSB exciter whose output sampling rate is $4\omega_0$. IFs between 9 and 100 kHz or more have been used successfully in both amateur and commercial designs. This architecture yields a great deal of flexibility to the designer. Changing sidebands is a simple matter of going from subtraction to addition in the summation preceding the DAC. Note that both addition and subtraction may be performed at that point to produce a double-sideband, suppressed-carrier (DSBSC) signal.

Separate information may even be placed on each sideband to produce

(continued on page 5-11)

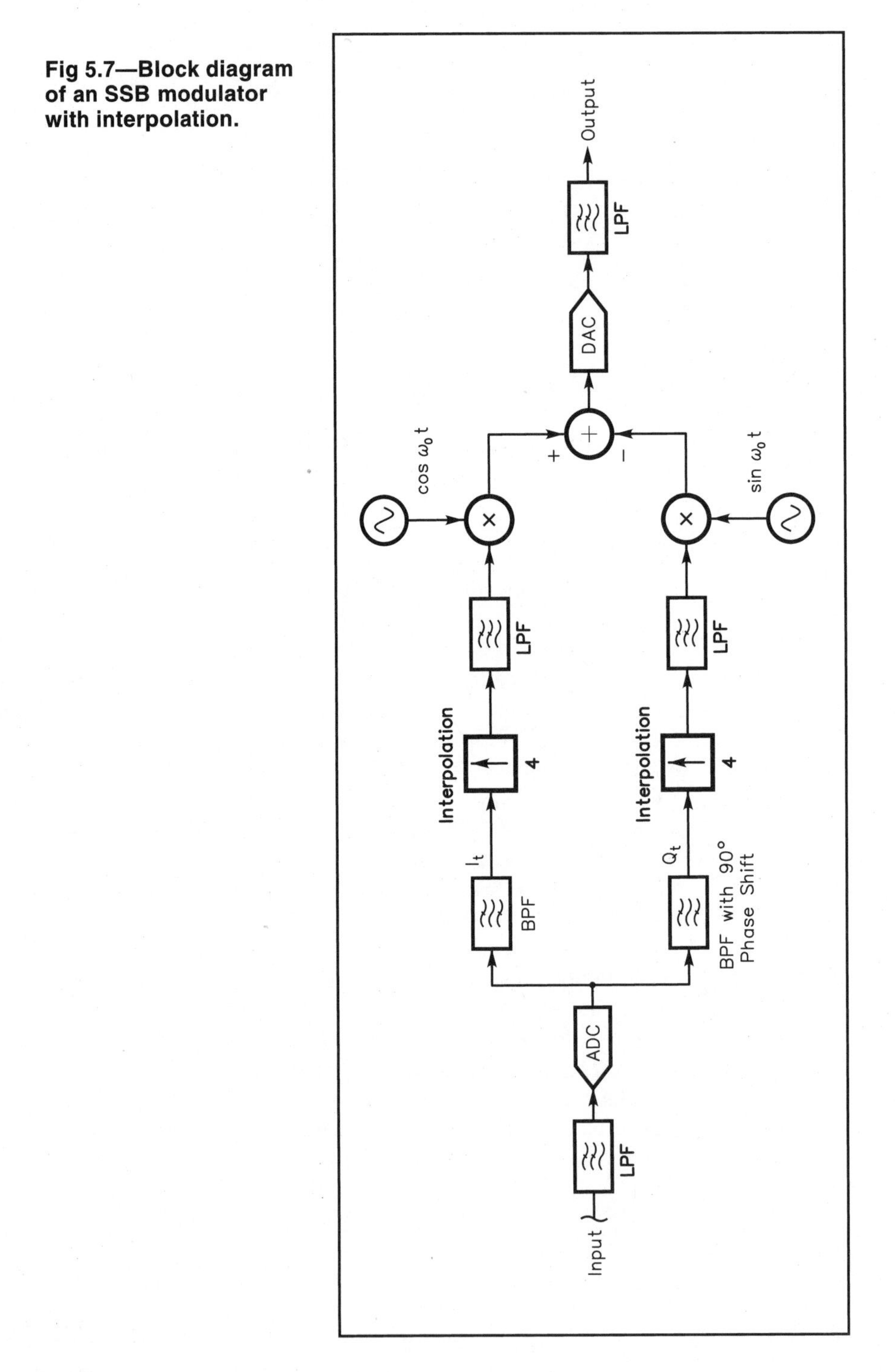

Fig 5.7—Block diagram of an SSB modulator with interpolation.

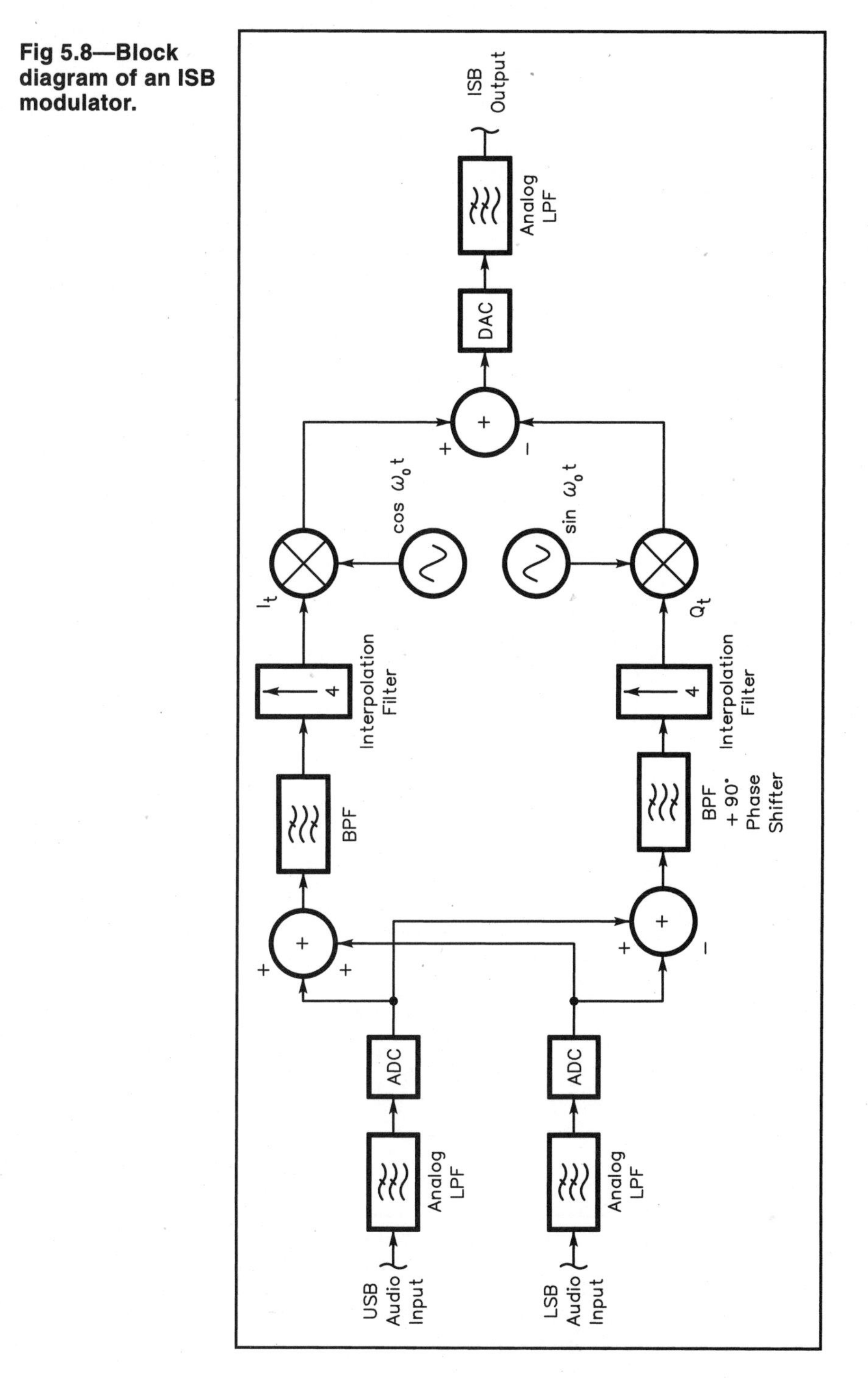

Fig 5.8—Block diagram of an ISB modulator.

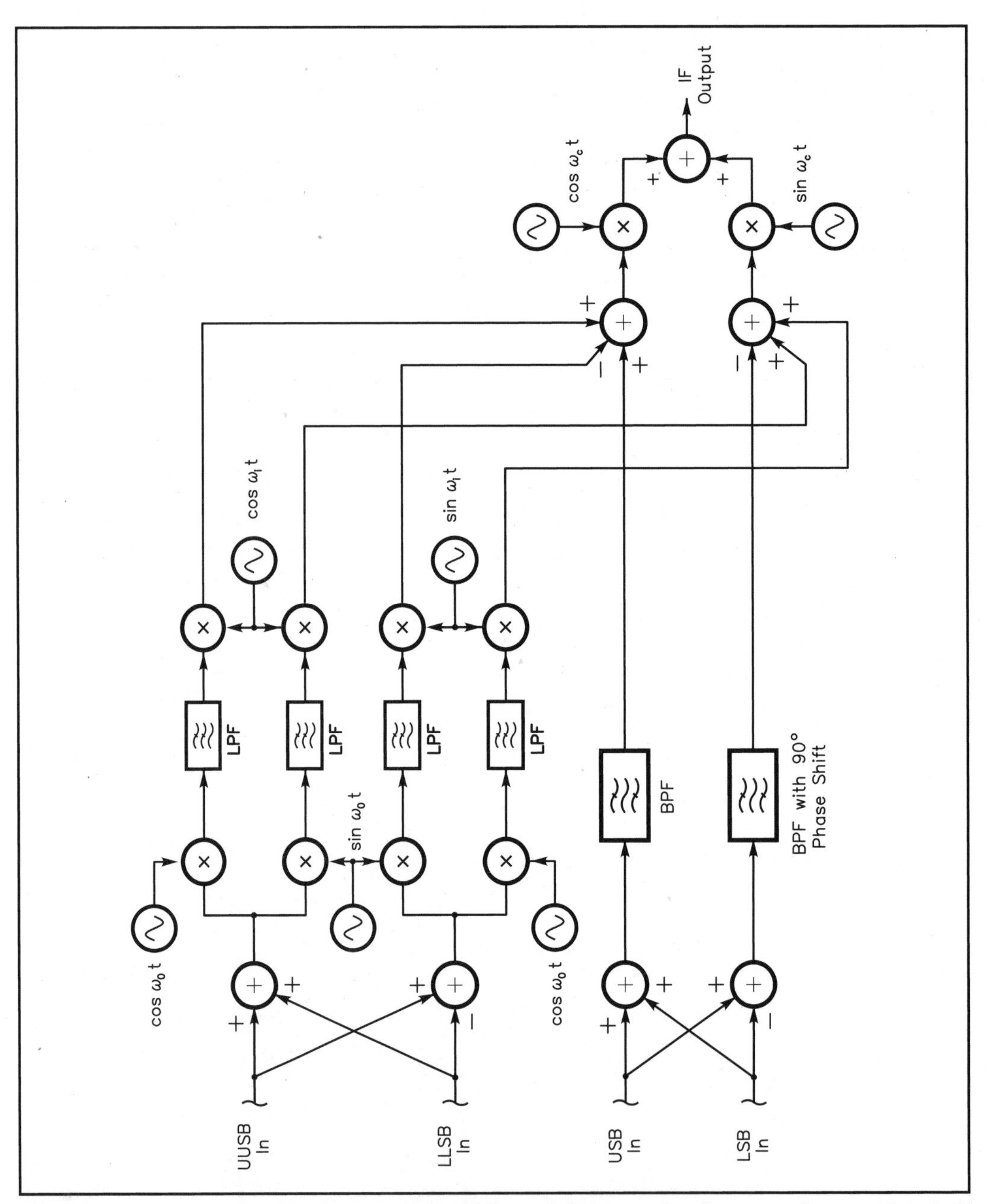

Fig 5.9—Block diagram of a four-channel ISB modulator. ω_o is chosen to be equal to half the channel spacing. ω_i is chosen to be equal to 1.5 times the channel spacing. The upper part of the diagram employs what is known as the *Weaver Method.* ω_c is equal to the final IF center frequency. Interpolators that are necessary to avoid aliasing along the way have been omitted for clarity.

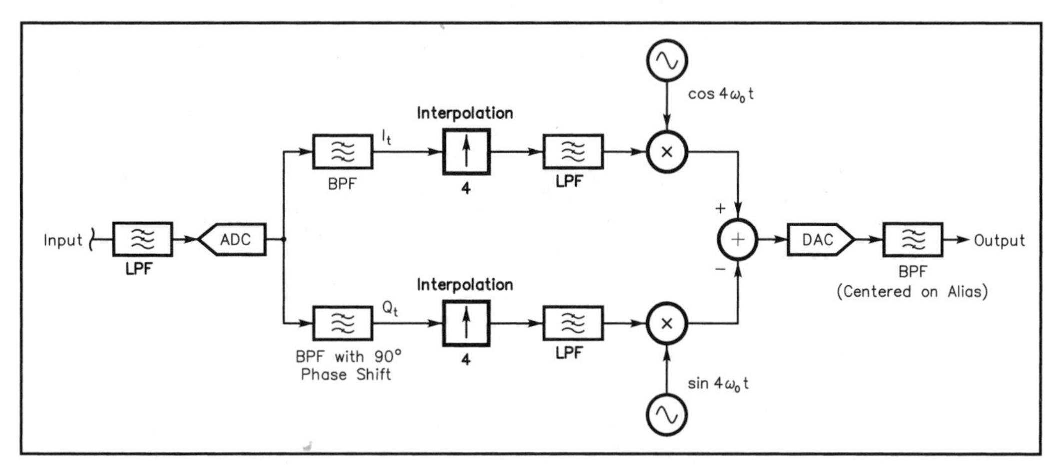

Fig 5.10—Block diagram of an SSB modulator using an alias, rather than the fundamental, as its output.

independent-sideband operation. See **Fig 5.8**. An additional ADC and anti-aliasing filter provide a second input to the system. Four or more signals may be combined in a similar fashion by replicating the circuit up to the interpolation filters and splitting the total bandwidth into segments using analytic filter pairs in different sub-bands. A common reason for wanting to do this is to accommodate many independent channels on a single transmitter. Such frequency division is shown in the diagram of **Fig 5.9**, which combines four signals.

The DAC at the output of Fig 5.9 produces a family of aliases at or near harmonics of the output sampling frequency that are usually eliminated by the analog LPF. Since the design uses an output sampling frequency of four times the IF, the nearest aliasing products are at least f_S / 4 away from the fundamental. That makes it easy to design an anti-aliasing LPF because its shape factor may be as high as two. By making that filter a band-pass, though, one of the aliases may form the output and the fundamental and other aliases may be rejected. This is an easy way to obtain an upward frequency translation and it might even eliminate mixers in some designs, as illustrated in **Fig 5.10**. The main trouble with that approach is that unless the DAC is rated for desired performance at the alias frequency, severe distortion may result. Also, alias amplitude decreases in a sin(x) / x fashion as alias number goes up. Therefore, aliases do not necessarily enjoy the full dynamic range of the DAC. Note that the bandwidth of aliases is identical to the bandwidth of their associated fundamental components. A frequency translation is obtained: Again, this is the result of a convolution in the frequency domain (the sampling process), which has been shown to be equivalent to multiplication in the time domain.

A Digital SSB Demodulator

Notice that in the modulator above, we filtered a signal, mixed it and so forth; but, for the most part, it retained its properties throughout the exercise. Now it follows that the process may be reversed and the signal translated back to a *baseband* or audio signal. The math works in one direction as well as the other.

The first job is to get some samples of a received IF signal. In the diagram of **Fig 5.11**, an analog band-pass filter precedes the ADC because harmonic sampling is being employed (Chapter 2). The IF where that is performed is largely determined by characteristics of available ADCs. Certainly prevalent at the time of this writing are low-frequency IFs and even some VLF. In addition, many cellular and satellite systems use much higher IFs and a lot of processing horsepower. When specifying ADCs, one should be interested in *dynamic range*: usually specified as *spurious-free dynamic range* (SFDR); and, of course, its sampling-frequency limit, which must be at least twice the bandwidth of interest.

Expertise in ADC technology is rapidly progressing. As it inches closer to the antenna, more traditional analog circuits will fall by the wayside. The trend these days seems to be toward higher levels of circuit integration—and of course speed, because that means bandwidth. How the bandwidth is sliced up depends only how received signals are processed.

After the data have been taken, they are mixed with a complex LO—usually to baseband, but sometimes to an intermediate IF. Again, it is well to choose

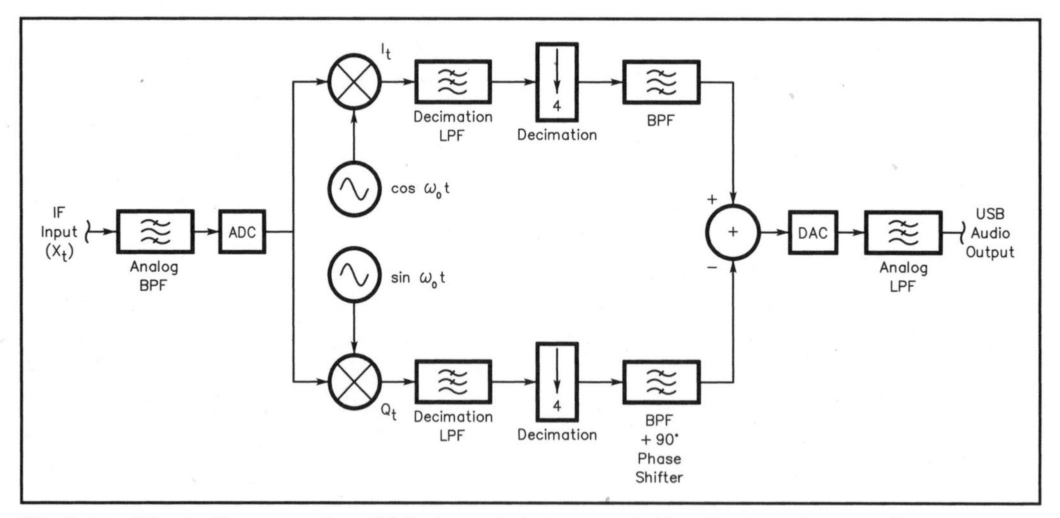

Fig 5.11—Block diagram of an SSB demodulator employing harmonic sampling.

an IF that is an integral multiple of the sampling frequency. That turns the mixing operation into simple *multiplexing*, just as in the modulator. Since the input and output signals are real, only two multiplications are necessary:

$$x_t e^{j\omega_0 t} = x_t \cos\omega_0 t + j x_t \sin\omega_0 t \qquad (50)$$

At this point, we are typically interested in one signal of many and we employ decimation to reduce the sampling frequency to twice the wanted bandwidth. We may combine decimation with low-pass or band-pass filtering as described in Chapter 2. The complex mix produces an analytic signal. This signal feeds an analytic filter pair that may be band-pass, thus determining final selectivity. The real part of the result is made in the summation: addition for LSB and subtraction for USB. An anti-aliasing filter again follows the DAC. The ISB receiver follows from this design. It obviously needs another DAC and LPF. And so it is for SSB demodulation: We call it a Hilbert *transform* because it only changes the representation of the signal, not necessarily the signal itself. That distinction also applies to Fourier transforms, treated in Chapter 8.

Properties of SSB Signals

In the discussion above, the peak amplitude of the LO is constant. The amplitude of an SSB signal may therefore be expressed as some function of the modulating signal. Thinking of the analytic microphone signal $I_t + jQ_t$ as a vector, its length is equal to its instantaneous amplitude:

$$A_t = \left(I_t^2 + Q_t^2\right)^{\frac{1}{2}} \qquad (51)$$

The phase of the signal is the instantaneous angle of that rotating vector:

$$\phi_t = \tan^{-1}\left(\frac{Q_t}{I_t}\right) \qquad (52)$$

Now we may write the real part of Eq 49 as[18]:

$$\Re\left[e^{j\omega_0 t}\left(I_t + jQ_t\right)\right] = A_t \cos\left(\omega_0 t + \phi_t\right) \qquad (53)$$

This shows an SSB signal to contain both amplitude and phase modulation: It is not just AM alone. Also, note that having defined the amplitude and phase of the baseband signal in Eqs 51 and 52 above, we may write:

$$\left(I_t + jQ_t\right) = A_t e^{j\phi_t} \qquad (54)$$

directly relating envelope and phase to the analytic baseband signal. Those properties of the baseband signal are identical to those of the transmitted SSB wave.

Let us examine how analytic signals lend themselves to handling other types of modulation, such as AM and FM. Those formats involve processes other than the simple frequency translation of SSB.

AM Demodulation

Despite being the oldest form of modulation, AM is particularly interesting in DSP implementations. In linear AM, the envelope of a signal is equal to the amplitude of the baseband information. A first try at AM demodulators utilized rectification and filtering, which were certainly simple methods that could be done with relatively few parts. A better way is to use the vector relationship of Eq 51, computing the envelope directly.

Now we are stuck with computing square roots; but, fortunately, a lot of work has been done on that subject over the centuries. In the 17th century, calculations had to be done entirely by hand. Anything that sped them along was a major blessing. Logarithms had only just recently been discovered. The logarithm tables of Napier and Briggs helped immensely, even in partial form, because of the relation:

$$x^{\frac{1}{2}} = \log^{-1}\left(\frac{\log x}{2}\right) \tag{55}$$

and because it is quick to divide by two.

Big logarithm books were hard to come by and other mathematicians, including Isaac Newton, sought other ways to find the roots of a function. Iterative methods were popular at the time: repetitive calculations of some function that converge on a result. It is interesting to note that this practice is still used in many DSP algorithms. Newton's method for square roots is as follows.

Take a crude guess at the square root of the *argument* in question. Divide the argument by the crude guess. Add the crude guess to this result and divide it all by two. Then, use this result as the new, slightly less-crude guess, repeating the process enough times to achieve the desired accuracy:

$$\begin{aligned} &\text{let GUESS}_{\text{new}} = \left(\frac{\dfrac{\text{Argument}}{\text{GUESS}_{\text{old}}} + \text{GUESS}_{\text{old}}}{2}\right) \\ &\text{let GUESS}_{\text{old}} = \text{GUESS}_{\text{new}} \\ &\text{REPEAT} \end{aligned} \tag{56}$$

In practice, the accuracy of this result reaches the limits of 16-bit, fixed-point representation within five or six iterations when the initial guess is good for the entire possible range of arguments. That is still a fair amount of computation and better results may be obtained in many cases by using interpolation, much as was done on sampled signals in the discussion above.

An Interpolation Algorithm for Square Roots

The temptation to use look-up tables—as did readers of Briggs' logarithms—in software to find square roots is an attractive one. Gigantic tables may not be much of a problem for chip manufacturers to implement, but they may tax available memory in embedded systems. One way to reduce the size of look-up tables is to use decimation and interpolation, wherein we may look between values in a table for the value corresponding closest to our desired argument.

To start, reduce the number of entries in a large table through the process of decimation. That is, simply eliminate all but every Nth sample, where N is the decimation ratio. A table with $2^{16} = 65{,}536$ entries, for example, may be reduced by a decimation factor of $N = 2^6 = 64$ to length of $2^{16}/2^6 = 2^{10} = 1024$.

In the interpolation algorithm, the two closest results corresponding to the argument are found in a look-up table having L = 1024 entries. A *first-order approximation* is taken to the square-root function between any two points: a straight line. The square-root function is assumed to increase linearly between table entries as the argument increases. This results in a linear, piece-wise depiction of the data, as shown in **Fig 5.12**. This is a very common construct in DSP. Although we do not have exact data about what the function does between samples, we may state that most of the information has been captured when the basic shape of the square-root function is retained.

Errors with this method produce distortion in the result that may be ana-

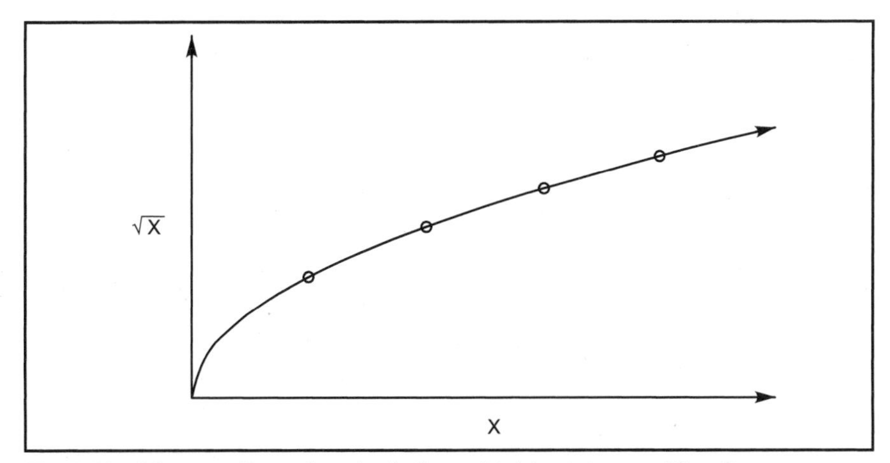

Fig 5.12—Linear, piecewise depiction of table data resulting from interpolation algorithm.

lyzed. An error function may be drawn, as in **Fig 5.13**, and its maximum found over the time interval for each sample. Such an error function is analogous to quantization noise, as outlined in the first chapter. This error function may be compared with the function $y_t = x_t^{1/2}$ to find the maximum error. That maximum error obviously occurs where the square-root function differs most from a straight line between entries. In Fig 5.13, two adjacent table entries are the square roots of numbers x and x + 1 / L. It is likely that the maximum error between the straight-line interpolation, AB, and the actual square-root function will occur near the middle of each line segment.

The point midway between the two end points, C, lies at:

$$y = \frac{x^{\frac{1}{2}} + \left(x + \frac{1}{L}\right)^{\frac{1}{2}}}{2} \tag{57}$$

The actual square root at the midway point is:

$$y = \left(x + \frac{1}{2L}\right)^{\frac{1}{2}} \tag{58}$$

so the maximum error is close to the difference of these vertical coordinates:

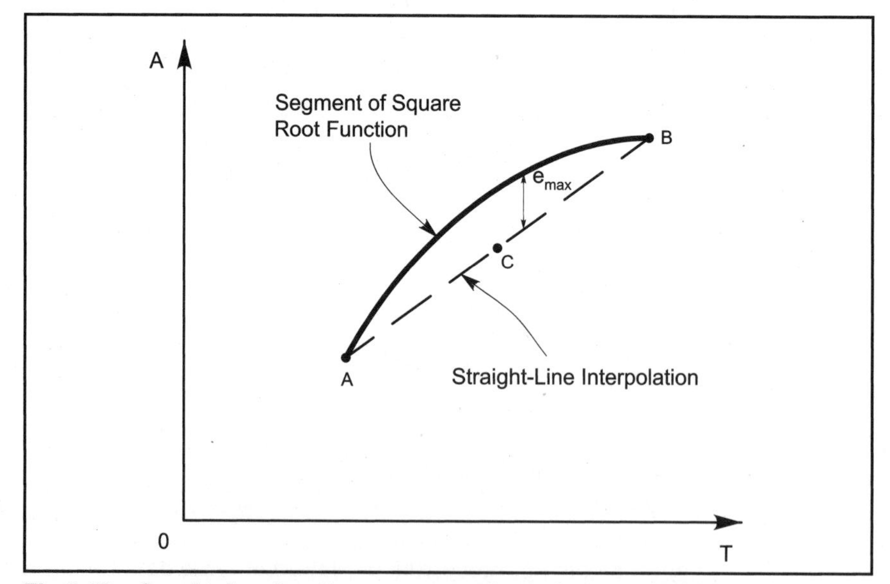

Fig 5.13—Graph showing the error magnitude between table entries.

$$e_{max} = \frac{x^{\frac{1}{2}} + \left(x + \frac{1}{L}\right)^{\frac{1}{2}}}{2} - \left(x + \frac{1}{2L}\right)^{\frac{1}{2}} \qquad (59)$$

It is perhaps surprising that the error shrinks with increasing x. For L = 1024, errors do not exceed 0.5% even for very small values of x; errors are vastly smaller for larger x. For example, when x = 0.3L, the error is approximately -3.5×10^{-5} %.

Sideband Diversity Reception

Linear AM poses some problems for relatively simple demodulators. Waves propagating via the ionosphere tend to suffer from *selective fading*: Amplitudes and phases of individual frequency components, including the carrier, may vary independently of others. When the carrier's amplitude is not constant relative to the modulation, a *carrier shift* is said to occur. When the carrier's amplitude drops, severe distortion is produced by demodulators employing envelope detectors, such as those described above. Also, energy in each sideband may not correlate properly, resulting in further distortion.

One way around these problems is to simply use an SSB demodulator to listen to one of the sidebands of an AM signal. While this obviously discards half the transmitted information, incurring a 3-dB penalty, that discarded information is redundant. The carrier contains no relevant information and may be eliminated by filtering in a receiver as well.

A system that has found some popularity among short-wave listeners is that of *sideband diversity*. In this system, the amplitude of each sideband is measured independently and the demodulator is automatically switched to the stronger of the two. It is remarkable that fading of two adjacent sidebands can occur at quite different times. Sideband diversity easily makes up the 3-dB penalty by improving signal-to-noise ratio during fades. The receiver's frequency accuracy must be good, or the listener will hear a difference when switching sidebands.

Synchronous and Exalted-Carrier Reception

To get the best frequency accuracy, we may phase-lock the receiver to the incoming carrier. DSP makes this especially easy to do, as shown in **Fig 5.14**. A narrow band-pass filter strips the modulation from the signal and squares (hard-limits) the remaining carrier, which acts as one input to a PLL. The other input to the PLL, the reference, is of course the receiver's center frequency. The PLL tunes the receiver to achieve lock. Capture range is obviously affected by the width of the band-pass filter. A neat feature of DSP implementations is that the filter's width may be altered to meet current loop conditions. When the PLL is

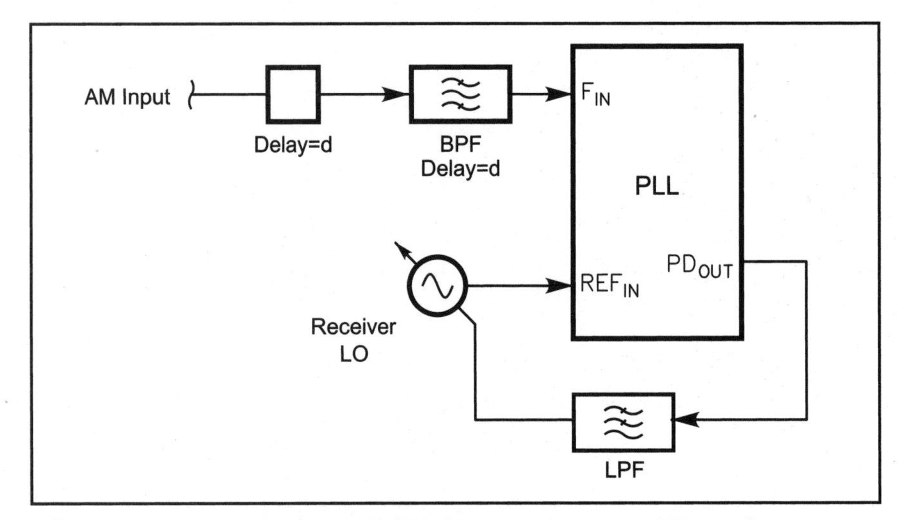

Fig 5.14—Carrier stripping and PLL in a synchronous AM receiver.

not locked, a wide filter may be employed to increase capture range; when lock is achieved, the filter may be narrowed until it excludes all modulation.

In many situations, *synchronous reception* gains up to 3 dB in SNR performance because it forces the phases of sideband components into their proper relationships. It also eliminates any frequency shift during sideband diversity operation, although in that case, the 3-dB improvement is not necessarily available. Finally, elements of a synchronous AM receiver lend themselves to *exalted-carrier* techniques that deal with fading of the carrier.

Ideally, a synchronous AM receiver contains a copy of the carrier from which all modulation has been stripped. That makes it possible to subtract the carrier from the received broadband signal to produce a DSBSC signal. The filtered carrier signal may be limited and added back to the DSBSC signal in its proper phase and at a constant amplitude. Alternatively, a locally generated copy of the carrier may be used since its frequency is exactly known. The regenerated AM signal then feeds a simple envelope detector as before. Carrier shift has been minimized and the advantages of synchronous demodulation have been retained. See **Fig 5.15**.

AM Modulators in DSP

Linear transmitters usually employ ALC to maintain constant peak-power output in the presence of modulation. Power amplifiers and other stages in a transmitter are rated by their peak-power capabilities and these ratings must not be exceeded, lest severe distortion be produced. It has long been a problem in AM transmitters to hold the carrier amplitude constant over the range of modulation percentages and, in the case of HF transmitters, over more than

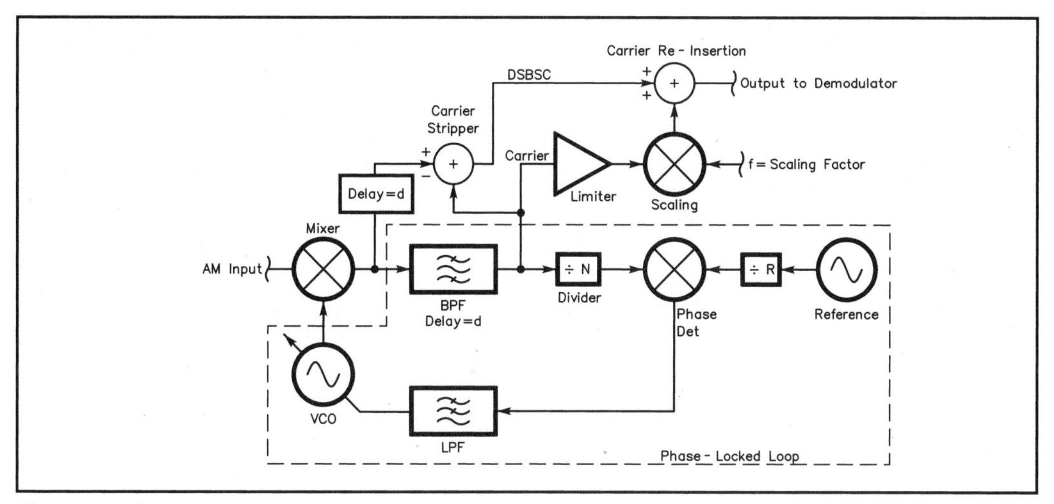

Fig 5.15—Exalted-carrier, synchronous AM demodulator in DSP.

four octaves of frequency range. Transmitter gain typically varies by several dB over that range. The peak envelope power (PEP) of a 100% modulated AM transmitter is four times the carrier power. At 100 W PEP, for example, we want the carrier to be a constant 25 W. Fixed carrier injection that achieves 25 W on one band might produce 50 W or more on another band where the transmitter gain is higher. In that case, the appearance of modulation might try to force 200 W of PEP, which is too high; ALC will reduce gain and therefore the amplitude of the carrier—carrier shift is the result. Carrier shift introduces transient distortion. Minimal though it might be, carrier-shift distortion in an AM transmitter may be avoided using adaptive DSP methods.

A DSP may readily compute the ratio of exciter drive level to transmitter output level when the transmitter is on. This information is used to find the drive level that reaches exactly 25% of the PEP setting. Now the baseband (audio) amplitude to the modulator must be held to a peak amplitude equal to the carrier's to maintain 100% modulation. That means that two servo mechanisms operate in this AM ALC circuit, shown in **Fig 5.16**. The first continually computes the drive-to-output ratio and sets the carrier power to 25% of the PEP setting. The second compresses the peak baseband signal to that same level and the result is always a 100% modulated AM signal.

Since the baseband peak detector employs a full-wave rectifier in software, baseband inputs with asymmetrical positive and negative excursions may produce unexpected results. To see this, one must understand the distinction between upward and downward AM modulation.[19] 100% downward modulation means the signal's amplitude is zero during the largest negative baseband excursion; 100% upward modulation implies signal amplitude of twice the

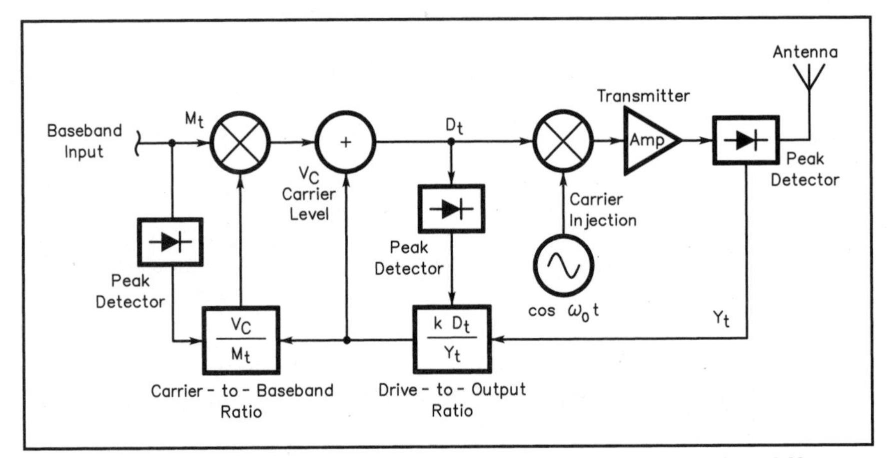

Fig 5.16—Block diagram of a digital AM ALC that avoids carrier shift.

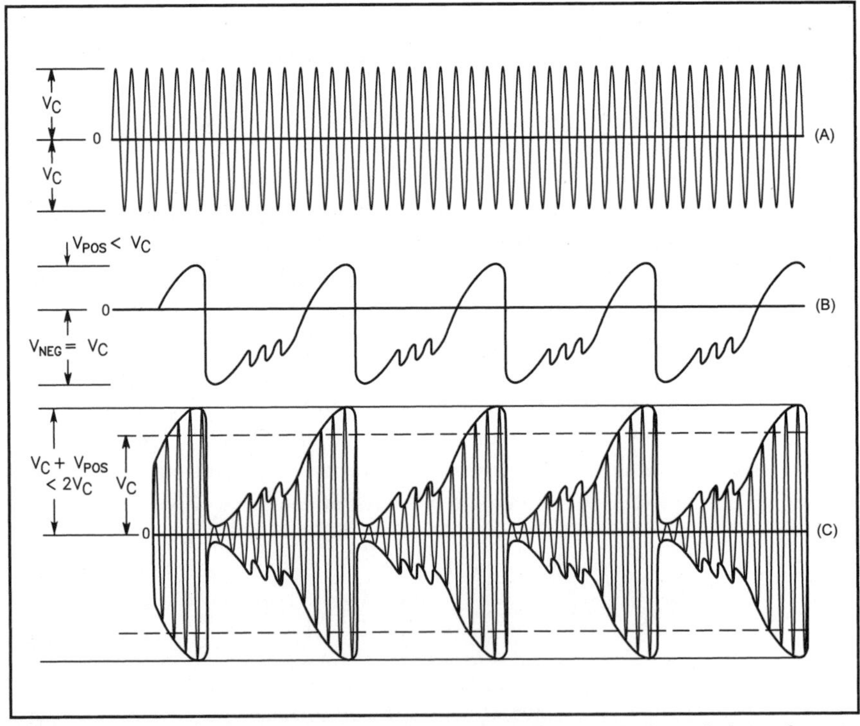

Fig 5.17—AM carrier, asymmetrical baseband input, and modulator input.

carrier's (four times the power) during the largest positive baseband excursion. In our DSP modulator, either the downward or upward modulation may reach the limit before the other can do so. If the downward modulation forces baseband compression first, the PEP will not reach its maximum level without introducing a carrier shift. See **Fig 5.17**.

Exponential Modulation (FM and PM)

In linear *exponential modulation*, the instantaneous frequency or phase of a carrier varies in direct proportion to the amplitude of the baseband signal. FM and PM are usually generated in DSP using a *numerically controlled oscillator*, which technique is identical to that of *direct digital synthesis (DDS)*. DDS and other synthesis methods are discussed in depth in the Chapter 7. Here, we shall focus on properties of exponentially modulated signals, especially those properties that are critical in DSP modulators and demodulators.

Let us begin with a simple, unity-amplitude cosine wave as our carrier: $\cos\omega_0 t$. To build an exponential modulator, it is once again convenient to employ analytic signals, so we add an imaginary part to the carrier: $j\sin\omega_0 t$. Now the carrier may be expressed through Euler's identity as:

$$\cos\omega_0 t + j\sin\omega_0 t = e^{j\omega_0 t} \qquad (60)$$

A Hilbert transformer is used to transform the real baseband signal into the analytic signal of Eq 54, this time with unity amplitude:

$$\frac{(I_t + jQ_t)}{A_t} = e^{j\phi_t} \qquad (61)$$

These two analytic signals are multiplied to produce an exponentially modulated analytic signal, only the real part of which is of interest:

$$\begin{aligned} \Re\left(e^{j\omega_0 t} e^{j\phi_t}\right) &= \Re\left[e^{j(\omega_0 t + \phi_t)}\right] \\ &= \cos\omega_0 t \cos\phi_t - \sin\omega_0 t \sin\phi_t \\ &= \frac{I_t \cos\omega_0 t - Q_t \sin\omega_0 t}{\left(I_t^2 + Q_t^2\right)^{1/2}} \end{aligned} \qquad (62)$$

where the script R indicates the real part. As mentioned before, this relation is not often used to build FM or PM exciters; but it is presented here to illustrate the flexibility of analytic-signal representation.

Note in the first line of Eq 62 that the argument $\omega_0 t + \phi_t$ represents a carrier having a phase offset somehow proportional to the modulation. This phase offset ϕ_t is proportional to the amplitude of the baseband signal, as noted at the start of this section. We have a PM signal whose instantaneous phase angle is equal to the argument. Now we wish to determine the *phase deviation* in radians so we can find the occupied bandwidth of the signal. Occupied bandwidth will be expressed in terms of frequency, so it would be nice to define the signal's instantaneous angular frequency from its phase.

Frequency is the rate of change of phase, so we may write:

$$\omega_t = \frac{d(\omega_0 t + \phi_t)}{dt} = \omega_0 + \frac{d\phi_t}{dt} \tag{63}$$

This means that in PM, the instantaneous frequency is proportional to the rate of change of phase of the modulation. Higher baseband frequencies produce higher *frequency deviation* than do lower frequencies. Naming the deviation $d\phi_t / dt = \Delta\omega$ allows us to define a *modulation index* that relates deviation to modulation frequency:

$$\beta = \frac{\Delta\omega}{\omega_m} \tag{64}$$

where ω_m is the angular frequency of the modulation. Now β is equal to the peak phase deviation in radians (Reference 9).

In straight FM, instantaneous frequency is directly proportional to baseband amplitude and deviation is not dependent on modulation frequency. For modulation by a single tone, $\cos\omega_m t$, and a predefined frequency deviation $d\omega$, we have:

$$\omega_t = \omega_0 + \Delta\omega\cos\omega_m t \tag{65}$$

and frequency deviation is independent of ω_m; but the peak phase deviation β is inversely proportional to ω_m.

Spectrum of FM and PM Waves

With knowledge of the peak frequency deviation, a first approximation to the bandwidth of an exponentially modulated signal is at least twice the frequency deviation, since both positive and negative excursions are expected. Unlike AM, though, sinusoidal FM and PM produce an infinite number of sidebands on both sides of the carrier that are separated by ω_m. Their amplitudes generally—but not always—decrease with increasing separation from the carrier and may be found using *Bessel functions*. A Bessel function translates modulation index β into a sideband amplitude (and phase); these functions are too complex to derive here, but graphs and tables of them are readily available.

The Bessel function for the carrier shows that its amplitude goes to zero when $\beta \approx 2.404$; this happens when frequency deviation is 2.404 kHz and modulation frequency is 1 kHz, for example. It is strange to see the carrier disappear on a spectrum analyzer, but any distortion present in the modulator produces a less-than-perfect *first carrier null.* In fact, the depth of the null makes a good indicator of the distortion level. In this example, many sidebands are prominent, as shown in **Fig 5.18**. It is obvious that occupied bandwidth is much greater than twice the frequency deviation.

When beginning to think about an FM or PM demodulator, one must be concerned with the distortion induced on a signal by limiting its bandwidth prior to demodulation. Elimination of higher-order sidebands certainly destroys information but modulation indexes may be chosen that minimize occupied bandwidth. For small phase deviation β, a PM wave may be approximated as in the aperture jitter discussion of Chapter 2 and only the first sidebands are significant. At modulation indexes less than one, high-order products fall off very rapidly. This is the reason 8-ary PSK (see below) and similar modes are so effective in reducing occupied bandwidth. According to Reference 19, necessary bandwidth may be estimated as twice the sum of the frequency deviation and twice the highest modulating frequency:

$$BW_{min} \approx 2\left(\Delta f + 2f_m\right) \tag{66}$$

In a typical case, Δf = 5 kHz and highest f_m = 3 kHz; Bw_{min} = 22 kHz. Actual practice accepts a bit more distortion than this and a 15-kHz passband is used.

One reason this is acceptable is that the human ear is not very sensitive to

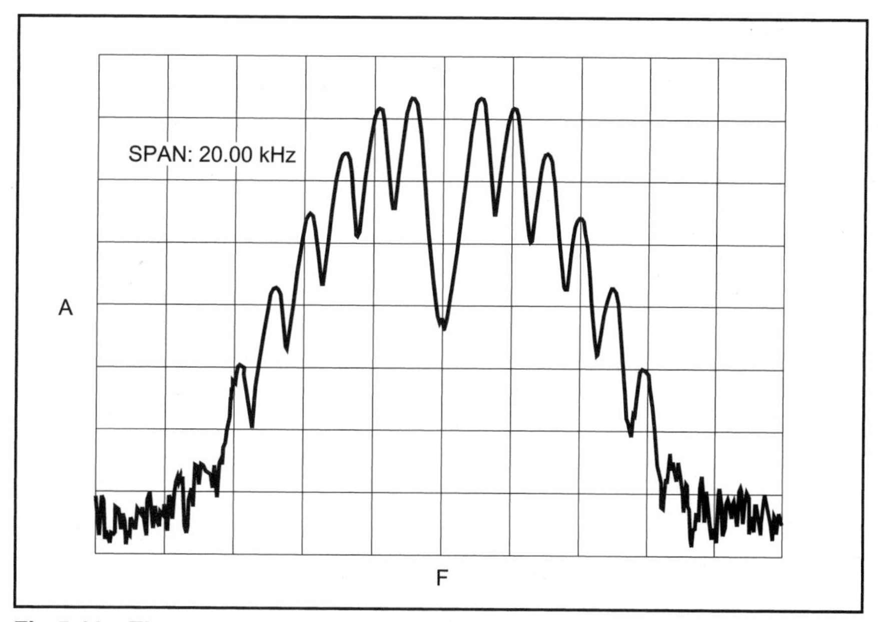

Fig 5.18—The spectrum of an FM signal at its first carrier null.

distortion in speech at high frequencies. Much of the energy in speech around 2-3 kHz is made up of plosive or fricative sounds like "p" or "f" that have little tonal content. That energy is very important to intelligibility, but is noisy and atonal, so distortion is less noticed. Inclusion of sidebands up to the second in this case is enough to preserve the intelligence in high-frequency baseband components for speech. Data applications may have to consider another set of criteria. Those are discussed later.

Exponential Demodulation

We have seen how beautifully analytic signal representations handle various types of modulation. Exponential or angular demodulation also benefits from this flexibility because, again, a modulated signal just carries the baseband information along for a ride; it does not alter it in any way, except for the bandwidth limitations noted above.

The discussion above began by showing that an exponentially modulated signal may be generated by multiplying two analytic signals: one is the carrier and one is the Hilbert-transformed baseband signal. To make the mathematics go in reverse, we might suspect that the inverse operation—division—may be used to recover the modulation. That is true and, following the notation of Eq 62, we may demodulate a real received signal x_t by first Hilbert transforming it:

$$x_t \xrightarrow{\text{Hilbert}} e^{j(\omega_0 t + \phi_t)} \tag{67}$$

and then dividing it by a complex carrier (BFO):

$$\begin{aligned} \frac{e^{j(\omega_0 t + \phi_t)}}{e^{j\omega_0 t}} &= e^{j(\omega_0 t + \phi_t)} e^{-j\omega_0 t} \\ &= e^{j\phi_t} \end{aligned} \tag{68}$$

Notice that the division turned into another multiplication, since:

$$e^{-j\omega_0 t} = \cos\omega_0 t - j\sin\omega_0 t \tag{69}$$

PM demodulators are not often built this way because other, simpler methods present themselves.

Demodulation by Finding Arctangents

Once a received signal has been Hilbert transformed, we may calculate its phase using the relation of Eq 52 above. The arctangents may be looked up from an interpolated table or they may be computed on the fly using the McLaurin series (first found by James Gregory, not McLaurin himself):

$$\tan^{-1} x = x - \frac{x^3}{3} + \frac{x^5}{5} - \frac{x^7}{7} + \cdots, |x| \le 1$$
$$= \sum_{k=0}^{\infty} \frac{(-1)^k x^{2k+1}}{2k+1} \tag{70}$$

Since a signal –I – jQ produces the same angle as signal I + jQ, *quadrant correction* must be applied by paying attention to the signs of I and Q. This amounts to adding or subtracting π depending on the initial result, or doing nothing.

Quadrature Detectors

An even simpler PM demodulator may be built along the lines of an old favorite in the analog world: the quadrature detector. In it, the received signal is multiplied with a delayed copy of itself. It might seem at first that this is a silly thing to do, but the time lag δ caused by the delay translates to a phase lag δω that varies with the frequency of the input, allowing recovery of the modulation:

$$\cos A \cos B = \frac{1}{2}\left[\cos(A+B) + \cos(A-B)\right]$$
$$\cos(\omega_0 t + \phi_t)\cos(\omega_0 t + \phi_t + \delta\omega) =$$
$$\frac{1}{2}\left[\cos(2\omega_0 t + 2\phi_t + \delta\omega) + \cos(-\delta\omega)\right] \tag{71}$$

Now the left-hand term in the result appears at twice the IF and is easily removed by filtering at the raw input sampling rate. The right-hand term has the information we want, but in the form of its cosine. When the delay is chosen to be an odd integral multiple of one quarter the period at the center carrier frequency, small frequency excursions about center still produce quite linear results. Near the zero-crossing, a cosine wave is nearly a straight line. An arccosine function and rotation de-normalization may be used to extend the linear range of the quadrature detector.

Also note that the equation above implies that the sensitivity of the detector increases with increasing δ. The longer the delay, the larger the output for a given deviation. The block diagram of a digital quadrature detector is shown in **Fig 5.19**.

Discriminators and Other Methods

Another method that has deep roots in the analog world is the discriminator. It is just a filter that has a linearly increasing (or decreasing) amplitude response versus frequency. Passing an FM signal through a discriminator

Fig 5.19—Block diagram of a quadrature detector.

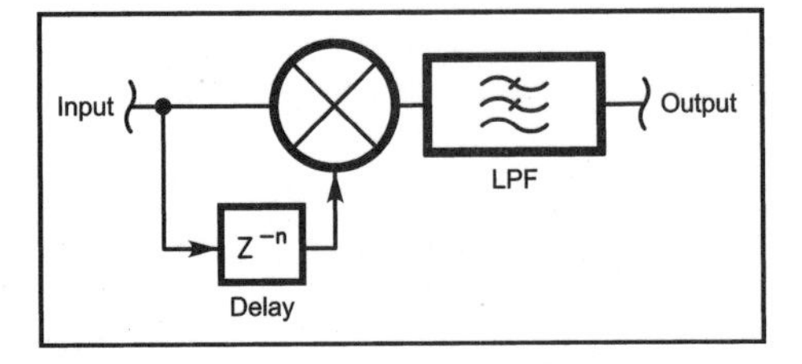

converts it to AM, which may then be demodulated by any of the methods outlined above for that mode. It is perhaps interesting to note that while the carrier amplitude might be zero in an FM signal, passing it through a discriminator results in an AM signal with normal, constant carrier amplitude.

Phase-locked loops have been popular as FM demodulators, and with good reason: They offer the possibility of *threshold extension*, or an increase in available signal near the noise floor. They are generally difficult to implement in DSP, but may operate with the same, familiar traits found in their analog brothers. One exception, of course, is that a discrete-time PLL only modifies its behavior at sample times, and not continuously. This becomes a factor when working close to half the sampling frequency.

Data Communications Formats

It seems that many traditional analog communications modes have gone over to digital coding. One of the first things that happens during any telephone call, for example, is that the audio is digitized and coded for further transmission. Why do we do this? Well, it is because digital data are robust; they only take on values of one or zero, thus detectors have a very clear-cut decision to make. It is easier to try to decide whether a received signal is a one or a zero than to decide exactly what voltage it is. Further, data may be regenerated at each point where they are detected. A digital repeater, for instance, may reform a data stream to perfectly comply with edge and clock times even though the stream was received with errors in those parameters. Finally, digital data may be coded in such a way as to detect and even correct errors that occur in transmission.

One of the earliest forms of digital transmission is CW—the on-off keying of a carrier and the earliest detector is the human ear-brain combination. That detection system is a darned good one and even contemporary engineers are hard-pressed to improve on it for CW. Surely, modern DSP filtering methods and entropy techniques do very well in aiding the listener, but in the finish, it is a human being who must make the decision as to whether what was received was real or an error.

CW is an excellent example of modes that shrink occupied bandwidth to the scale of the information rate. A proficient operator may send and receive 30 wpm CW, which is equivalent to about 20 bits per second (bps). When a CW character is four elements (dots or dashes) long, and counting the interelement spaces as bits, each character requires about eight bits—just the same as most text-coding schemes such as ASCII. Many operators can type a lot faster than they can send CW, though, so the efficiency of the coding scheme is lost because the human interface limits performance. Machine-generated and -detected CW can go one heck of a lot faster, but it is not necessarily the most bandwidth-efficient arrangement under those conditions.

With proper envelope shaping, CW occupies bandwidth proportional to speed. Envelope shapes that minimize bandwidth, though, are not necessarily those that aid in high-speed copying. Many operators find that sharper rise and fall times, with the associated increase in bandwidth, are easier to copy at high speed. These effects and others associated with detection theory are not completely understood and constitute an area of ongoing research.

It was obvious even to early experimenters that on-off keying is not the most advantageous means of transmitting text over radio. For one thing, on-off keying may suffer from the ill that when the carrier is off, interference may jump in and fill the void, destroying information. For another, much of natural and man-made noise has a large AM component. Both these issues are addressed by using frequency diversity in the form of *frequency-shift keying (FSK).*

Binary FSK

By frequency diversity, we mean that the energy contained in a signal is shared redundantly by more than one channel, if you will. Were the information in one channel suddenly removed, it could still be recovered from the other, unimpaired channel. That is the case for binary FSK, which uses two frequencies. One frequency represents a one or *mark*, the other a zero or *space*. A transmitted signal's frequency is made to jump from one frequency to the other in response to a digital data input. Each frequency may be analyzed separately and each looks like an on-off keying waveform. It follows that when one tone is on, the other must be off, and vice versa. Good FSK demodulators exploit this fact to advantage. It is a notable advantage because selective fading on HF circuits may null one tone while the other is copyable.

Binary FSK lends itself to spectral analysis by thinking of it as two on-off-keyed waves. We may also analyze it as an FM signal having pseudo-square-wave modulation. Certainly, the spectrum of an FSK signal is the convolution of the spectrum of the input signal and that of the carrier. Note that sidebands appear around each tone frequency at separations of roughly $^1/_2$ the baud rate. The *shift*, or frequency separation between tones, must therefore be at least equal to the baud rate to avoid overlapping the first set of sidebands. In practice, larger shifts achieve superior results with most demodulator designs.

As an example, take a common situation on the HF packet sub-bands: 300-baud FSK with a 200-Hz shift. The first sideband from the high-frequency tone lands within 50 Hz of the low-frequency tone, and vice versa. This makes it much more difficult to get some of the detectors discussed below to function properly. Some of the typically high error rates with this mode are caused by ionospheric distortion of signals, but it does not help to have your sidebands overlapping. That smears the data and a demodulator's decision is no longer clear-cut.

FSK RTTY using the Baudot code was the prevalent mode on HF for what seems like eons, and it lives still! Audio FSK signals are reasonably easy to generate using DDS techniques in software as described in the Chapter 7. Let us focus here on demodulation methods for binary FSK.

FSK Demodulators

For FSK detection, many designers find that some traditional analog methods are equally applicable to DSP implementation. Several of them, though, result in quite complex structures that are not often used in DSP. Among those are PLL methods; they will not be discussed here.

An attractive method for non-coherent demodulation involves energy detectors centered on each of the tone frequencies.[20] Band-pass filters are employed to segregate the two on-off signals and a vote is taken at each sample time on which channel contains the highest energy level. A block diagram of such a system is shown in **Fig 5.20**. This system produces good results even during deep selective fading since it is capable of making a very fine decision during the vote. Even when one of the tones is absent, the thing still copies the data because it has a variable threshold at the decision point by virtue of the fast-attack, slow-decay post-detection filters.

Treating an FSK signal as FM opens the door to the use of any of the FM demodulators described above. Quadrature and discriminator types have been used with success, followed by a threshold detector. Those FM demodulators may have a slight SNR-performance advantage over incoherent detectors and may be constructed with a single filter. The block diagram of a DSP discriminator using analytic signals is shown in **Fig 5.21**. As in linear FM demodulators, the sampling frequency must be high relative to the sampled signal's frequency to avoid running into non-linear effects that might find their ways into one's passband of interest.

Design of filters for FSK demodulators must take into consideration the spectral characteristics of the transmitted signal and the desire to optimize bandwidth for best performance near the noise floor. In general, the best channel filters have bandwidth that is comparable with the baud rate. As noted, this allows passage of the first set of keying sidebands and little else. An FSK signal detected at that bandwidth more resembles a sine wave that a square wave; post-detection filtering and threshold detection restore the data to their proper shape. The narrowest-possible filters ensure the best-possible SNR performance. Additionally, if a transmitter does not employ *phase-contiguous keying*, a

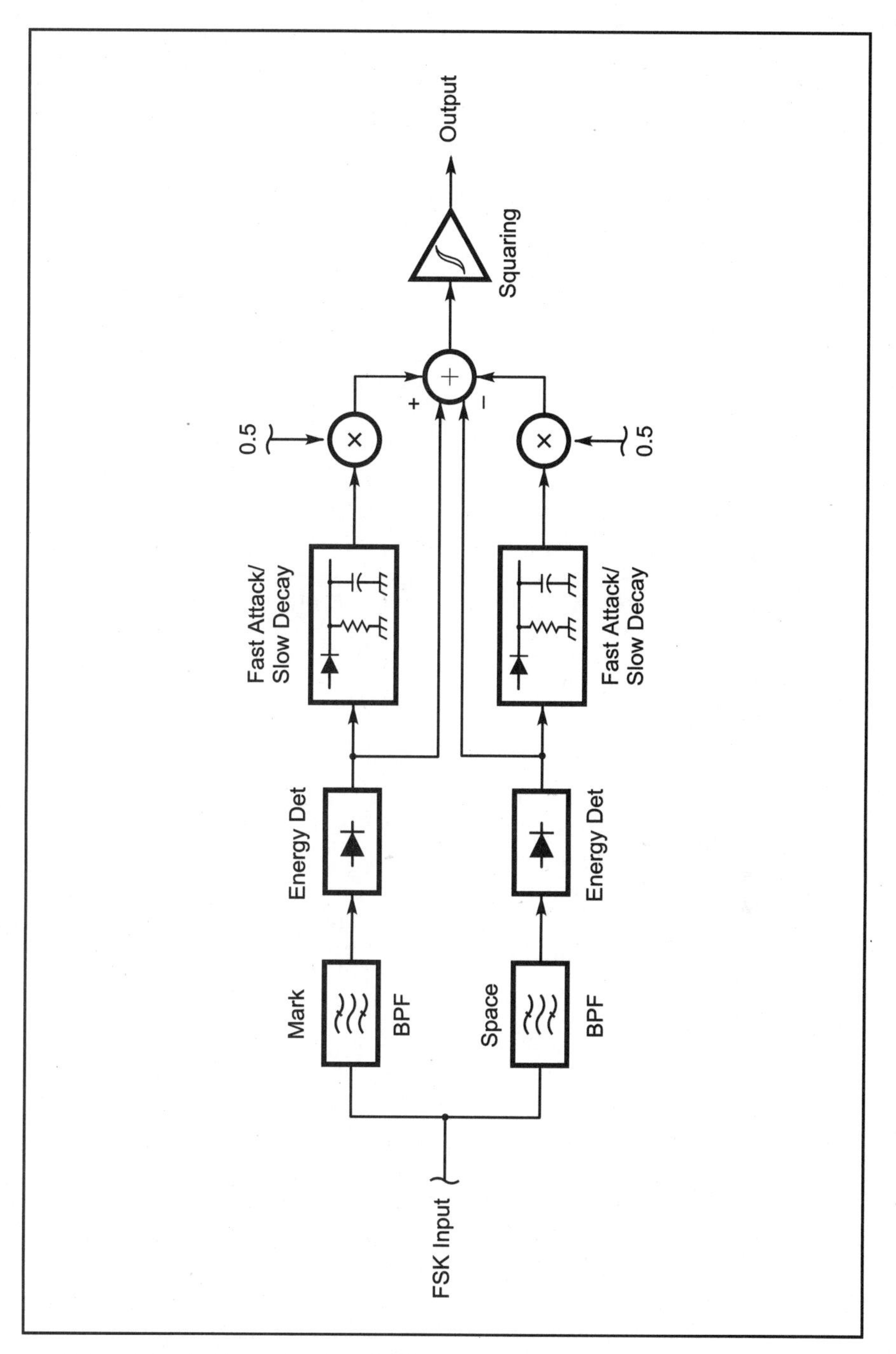

Fig 5.20—Block diagram of an FSK demodulator with differential energy detection.

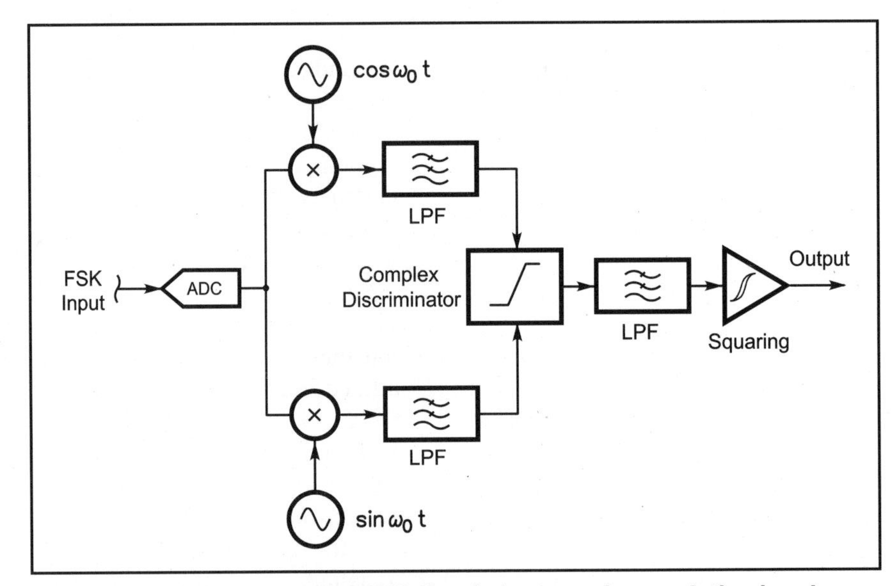

Fig 5.21—A digital, complex FSK discriminator using analytic signals.

demodulator's job is made more difficult because the energy is spread out more than necessary.

It is interesting to note that DSP versions of traditional FSK demodulators do not necessarily achieve significantly better results than their analog counterparts, because very narrow filters and extremely accurate phase control are neither needed nor desired. Even at 1200 baud, the group delay of analog IF filters is not usually a concern; however, FSK may be used with more than two tones to boost throughput and in this case, DSP stands out by eliminating a lot more hardware than in binary FSK.

M-ary FSK

In binary FSK, each tone represents a single bit. Now let us expand the number of tones to some number M, an integer greater than two. M is usually chosen to be close to an integral multiple of two, such as 4 or 8. When each tone is FSK modulated with baud rate R, the total throughput is MR. Occupied bandwidth obviously increases, but we are passing more data. Each tone still represents a single bit, but M bits are being transferred at each sample time.

Another way to manipulate M tones is to make each represent more than one bit. In a system using 16 tones, for example, each may be taken to represent $\log_2 16 = 4$ bits. In the modulator, only one of the tones is on at any one time. The transmitter's output hops from one frequency to the next based on groups of four bits at its input. A 16-tone system built that way has about the same occupied bandwidth as a 16-tone system in which each tone is FSK modulated; but it requires only 16 filters instead of 32 and the amplitude and frequency offset of keying sidebands is reduced. That means the pre-detection filters may be narrower, achieving an SNR increase.

DSP's ability to perform a spectral analysis of input signals is a valuable commodity when it is time to demodulate some M-ary FSK signals. Rather than building the 16 filters and energy detectors of Fig 5.21, we may employ Fourier-transform techniques to build a simpler demodulator. A detailed explanation of Fourier transforms is presented in Chapter 8, but the basic principles are applied here to simplify the circuit.

Embarking on the job of building frequency-selective energy detectors, such as are needed, you will note that a very simple process translates a sampled signal to analytic form, as outlined near the start of this chapter. It yields two number sequences that relate the amplitude and phase of the signal at the analysis frequency. It is very much like building a spectrum analyzer on a budget.

Multiplying a sampled data sequence by a complex oscillator $\cos\omega_0 t + j\sin\omega_0 t$ mixes it downward and upward by frequency ω_0. Components in the sampled signal at frequency ω_0 naturally appear at zero frequency—they make a dc component. This dc component may be extracted by integrating (summing) all the samples of the mixed signal. Recall that our shorthand for the oscillator wave is:

$$e^{j\omega_0 t} \tag{72}$$

and we will call the sampled signal x_t. Multiplying sample-by-sample and summing, we have:

$$\begin{aligned} X_t &= \sum_{k=0}^{L-1} x_t e^{j\omega_0 t} \\ &= \sum_{k=0}^{L-1} x_t \cos\omega_0 t + j\sum_{k=0}^{L-1} x_t \sin\omega_0 t \end{aligned} \tag{73}$$

A multitone signal is shown in **Fig 5.22** being multiplied by the real part of the complex oscillator or $\cos\omega_0 t$. This seems a simple, yet time-consuming operation but the dc component of the product is evident in the result: It rides way above the zero line. Summing all the samples in the result finds the dc component and this is the real part of the analytic signal. Compute the sum of the sine product to get the imaginary part of the analytic signal.

Two multiplications must be performed for each sample and we have L samples. For L equal to the length of pre-detection filters above, though, computational burden is the same. Eq 73 is, in fact, a Fourier transform. While it is not optimized for purposes of demodulation, it allows detection of narrowband energy, producing an analytic signal that has amplitude and phase defined by Eqs 51 and 52, above. We are mainly interested in the amplitude at this point

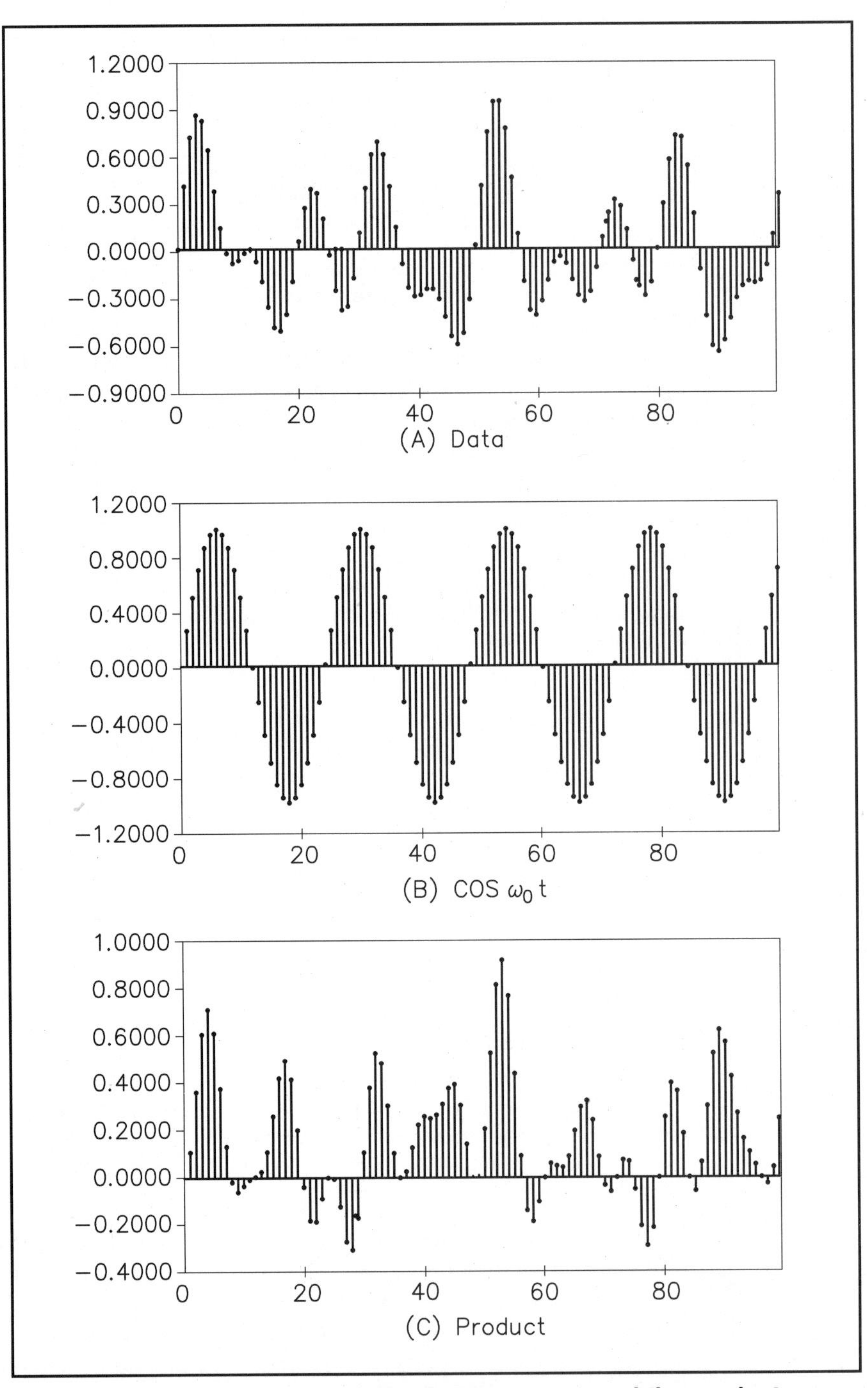

Fig 5.22—A multitone signal, multiplied by cosω_0t, and the product.

and an envelope-detector algorithm after Eq 51 is generally used to find it. Readers may be wondering how the same process may be called both a Hilbert transform and a Fourier transform. Little distinction really exists and they are basically the same thing. What may be said about the difference is that Fourier transforms aim to analyze frequency content and Hilbert transforms aim to produce analytic signals for mixing, modulation and demodulation purposes.

Binary PSK

PSK is conservative of bandwidth and that is an outstanding reason for its popularity, especially for synchronous data transmission. In *binary PSK*, the phase of the carrier depends on the modulator's input state. When the input changes state, the phase of the carrier changes by π radians; that is, the wave inverts its polarity. This process may be illustrated as in **Fig 5.23** by a multiplier whose inputs are the carrier and either 1 or –1. The output may be translated upward in frequency for transmission over radio or used as-is on a telephone line.[21]

The envelope of a bandwidth-limited BPSK signal is not constant since it reaches zero during phase transitions. It therefore requires linear signal processing in analog stages of a transceiver where bandwidth is to be conserved. See **Fig 5.24** for an example of a bandwidth-limited PSK signal at RF.

Note that bandwidth-limited BPSK resembles two-tone, SSB modulation when continual phase-reversals take place.[22] In that case, a PSK signal is composed virtually of two tones separated by the baud rate. Also note that this implies an average transmitted power of $^1/_2$ the PEP rating of the transmitter. 3 dB are lost from the equivalent FSK signal, but they are gained back in the reduction of occupied bandwidth.

Of interest is how much bandwidth PSK occupies relative to FSK at the same baud rate. Now the modulation index, β, of exponential modulation such as this is just equal to the peak phase deviation in radians. The total phase swing for BPSK is π radians and so its peak value about zero is π / 2 radians.

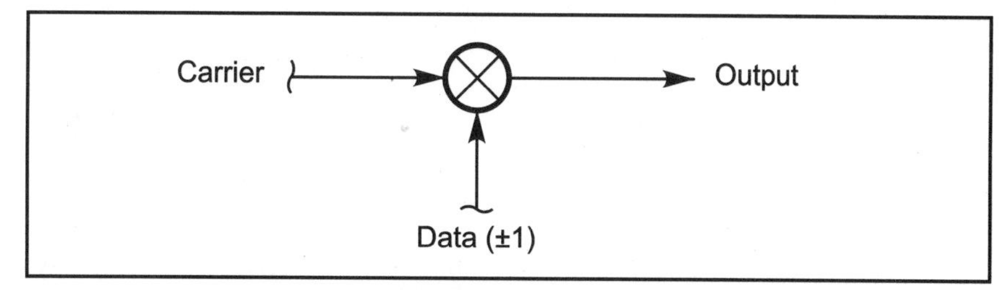

Fig 5.23—Block diagram of a BPSK modulator.

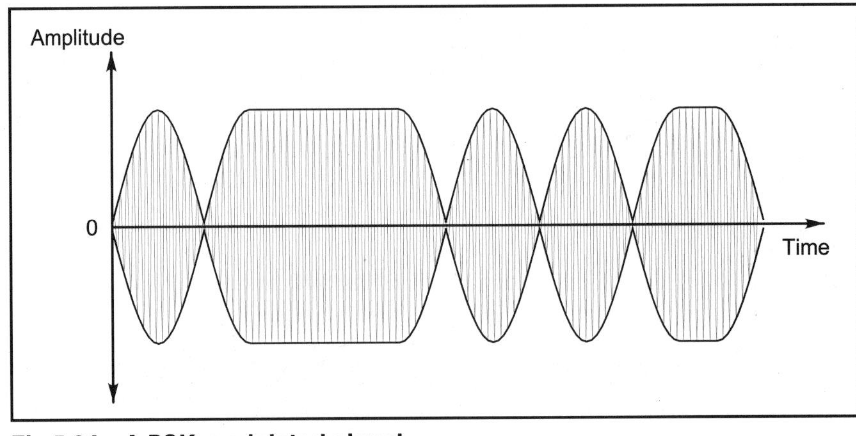

Fig 5.24—A PSK-modulated signal.

This is about 1.57 and first-sideband amplitudes may be computed via the Bessel functions mentioned before. Third and higher odd-order sidebands are accentuated by square-wave modulation. Do not forget that square waves have lots of odd harmonics and those must be included in the analysis. If not mitigated, they tend to make FSK and PSK signals quite broad. Subsequent filtering is usually required to meet commercial emissions limitations in most countries. That means, again, that we are going to have to chop off some high-order sidebands of the transmitted signal, and maybe of the received signal as well.

Just as for FSK, at least the first sidebands must be included to get even a sine-wave approximation to the data, and that means BPSK's significant bandwidth is about the same as FSK's. That PSK's high-order sideband amplitudes are lower, though, means that fewer of them need to be included to accurately reproduce the modulation. In other words, the modulation index is lower and performance depends less on high-order sideband energy. Compare BPSK's modulation index of $\beta \approx 1.57$ to that of a typical FSK signal with $\Delta f = 85$ Hz and f = (baud rate) / 2 = 22.75 bps (60-wpm RTTY) having $\beta = \Delta f / f = 85 / 22.75 \approx 3.74$ and you will see the benefits of PSK. Higher modulation indexes mean more occupied bandwidth.

An interesting result occurs when a received BPSK signal is squared. Half-cycle phase flips then become full-cycle, which means the second harmonic of the carrier wave is thereby recovered from the signal. That characteristic may be of importance in many applications, because it implies that synchronous detection is readily possible and computational complexity may be reduced.

It makes sense to shrink occupied bandwidth to the scale of twice the baud rate. PSK31 and similar modes have brought that out nicely. Would that we could do the same for phone emissions, the information in which is sometimes—

well, not much. As noted in the general FM case above, modulation indexes less than unity reduce sideband amplitude still further. It is not too hard to see that by making carrier phase shift less than one radian, more bandwidth reduction is achievable and additional possibilities for encoding arise.

PSK31

PSK31 is a sort of hybrid system in that it transmits a phase reversal when a zero is input; a binary one results in no phase reversal. This is a popular arrangement since it obviates the need to recreate a coherent phase reference at the receiver. **Fig 5.25** depicts a PSK31 demodulator circuit. Pulse shaping is employed in the diagram using a pre-modulation low-pass filter. The popular amateur RTTY mode PSK31 uses *differential BPSK (DPSK).*

To demodulate DPSK, we may compare the phase at the current bit time with that of the previous bit time. That is accomplished by something that looks a lot like a quadrature detector: **Fig 5.26A**. The received signal is multiplied by a one-bit-time-delayed copy of itself, followed by post-detection filtering and threshold determination. As in the case of BPSK, the bit clock may be recovered from the received data by detecting the received signal's envelope (with an AM detector), then using the envelope to drive one input of a PLL. In the locked condition, the PLL's VCO outputs a copy of the bit clock. See Fig 5.26B.

To make this scheme work, a preamble consisting of a continuous string of zeros is usually transmitted prior to letting the actual data fly. The repeating phase reversals allow the bit-clock PLL at the receiver to get locked before real data come along. During data transmission, long strings of ones would let the PLL drift away from the correct bit clock. Since no phase reversals would be transmitted, no information would be available at the receiver about when to

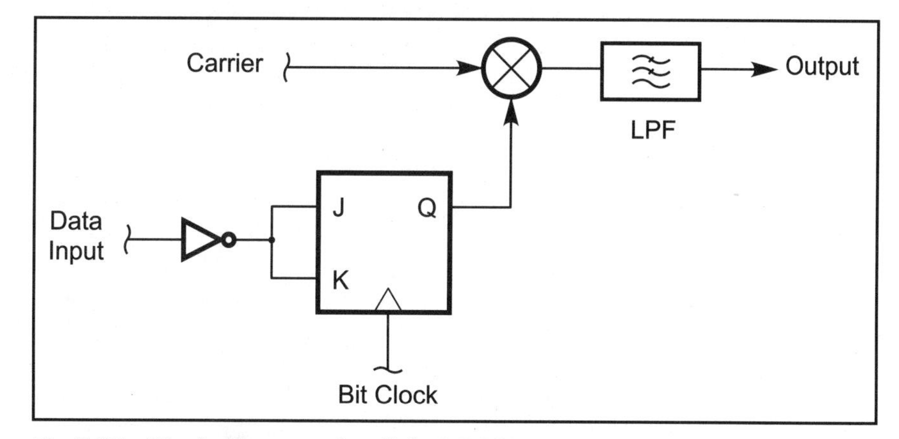

Fig 5.25—Block diagram of a digital, DPSK modulator.

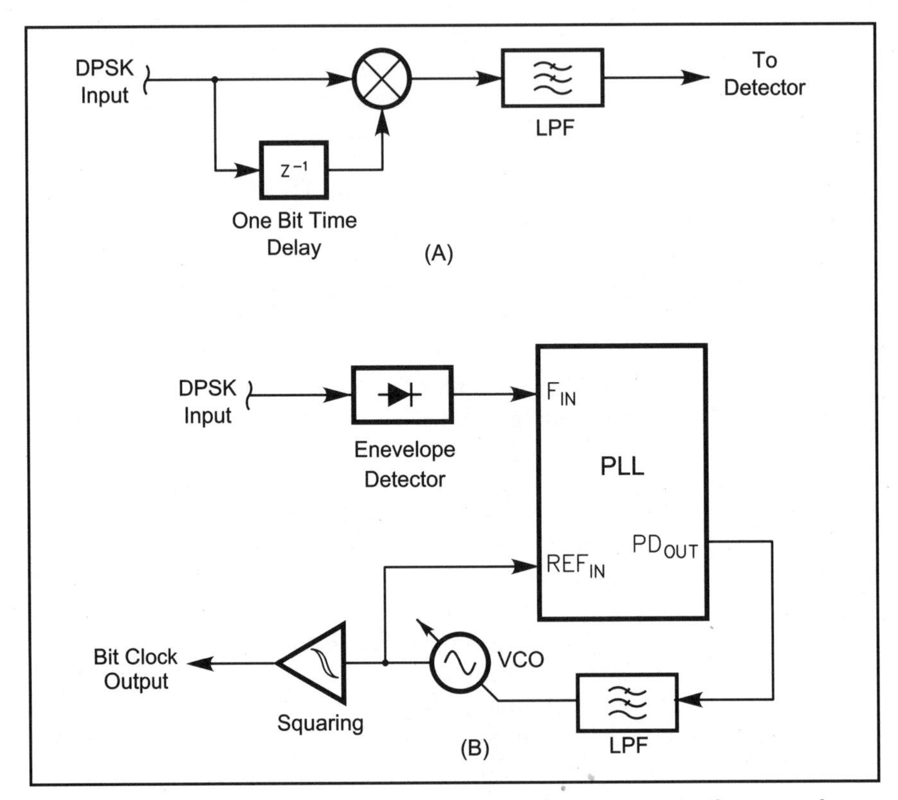

Fig 5.26—At A, a digital, DPSK demodulator. At B, a block diagram of a clock-recovery circuit.

expect the next bit time. PSK31's idle signal is a continuous string of zeros and so the system re-synchronizes very accurately when traffic is not being sent.

Note that it is also possible to compare a received signal's phase during the two halves of each bit time.[23] To do this, the one-bit-time delay of Fig 5.26A is replaced with a one-half-bit-time delay. An even higher sampling rate may be used in a DPSK demodulator to take many phase samples during a single bit time and a majority vote taken as to the phase. The occupied bandwidth of a DPSK signal is the same as that of the equivalent BPSK signal. For PSK31, the bit rate is 31.25 bps and so a band-pass filter producing optimal SNR performance must be at least 31.25 Hz wide to pass the first set of PM sidebands. A post-detection filter may be low-pass with a bandwidth of at least half the bit rate or 15.625 Hz. Such filters are relatively easy to design in DSP.

DPSK carries the same kind of disadvantage as BPSK: Phase reversals cause the carrier's amplitude to pass through zero and a continuous string of phase reversals results in a 3-dB decrease in average output power. During continuous transmission of ones, though, output is a continuous, unmodulated wave. This discrepancy in output power between states may cause problems in some systems by affecting SNR.

QPSK and M-ary PSK

Quadrature phase-shift keying, or *QPSK*, reaches a peak phase deviation of $\pi / 4$ or about $\beta = 0.785$. At this level, virtually only the first sidebands are significant and bandwidth is reduced to nearly the information rate. Each transmitted phase of 0, $\pi / 2$, π, and $3 \pi / 2$ radians represents two bits. The modulation index has been reduced by two, which means sideband amplitudes will drop by at least that much. Now, at least twice the data may be carried in the same bandwidth as before.

Multiple PSK waves may be combined as in the FSK case above to produce *M-ary PSK*. Again, we have choice of replicating PSK-modulated waves across the occupied bandwidth or letting each phase represent more than one bit. Quite often, a compromise is selected and multiple QPSK signals are combined within some target bandwidth. PM may even be combined with AM to exploit the advantages of each. An example of this is given below; but first, we will revisit the business of occupied bandwidth of PSK signals to see how it may be reduced to the minimum.

What we are learning is that, just like the rise and fall times of a CW signal, sharp transitions increase sideband energy in FSK/PSK systems. New approaches have now been taken that minimize the broadening of occupied spectrum by the transitions. Surely, we may just modulate without shaping, but subsequent filtering is necessary to meet most requirements. In some cases, bandwidth is getting so small that it is easier to do the shaping than to filter.

Minimum-Shift Keying

In *minimum-shift keying (MSK)*, each QPSK pulse is shaped by envelopes of sinusoidal form.[24] Researchers have shown that MSK is equivalent to FSK with a deviation of one quarter the baud rate, or $\beta \approx 0.5$. This is getting pretty low and the process approaches linearity.

Gaussian minimum-shift keying (GMSK) takes the shaping process a step further by finding the shape that minimizes occupied bandwidth. Carl Gauss was unequaled by any mathematician of his time. His *divergence theorem* forms the basis for modern field theory. His accomplishments are astonishing and include a very accurate version of celestial mechanics. He also blazed trails into non-Euclidian geometry that proved indispensable to Lorentz, Poincaré, and Einstein. Gauss found the energy distribution that corresponds to minimal bandwidth for an exponentially modulated signal. This distribution is easily transformed to the time domain by a Fourier transform. That is the rise and fall shape that satisfies the condition of minimum bandwidth.

Hybrids: Quadrature Amplitude Modulation

Evident from the analytic representations above is that we may exercise both amplitude and phase of a wave to convey information. To bring in an AM component, *quadrature amplitude modulation (QAM)* is used in contemporary systems such as telephone, satellite modems and digital voice systems to reach high data rates. Shannon[25] informs us, though, that to get those rates, SNR must meet a certain minimum. In addition, group-delay and quantization effects must be low. The group-delay requirements are not as severe as for QPSK, for example, and so QAM is well-suited for telephone lines. AM sidebands are generally less complex than PM sidebands and this fact has contributed to the popularity of QAM.

In one QAM system, four discrete amplitudes and four discrete phases are employed to allow a total of 16 discrete states. For an analytic signal, this is the same as four discrete amplitudes each for both I and Q. This shows that such a QAM signal is the same as the superposition of four *quadrature amplitude-shift keying (QASK)* signals. Each of the 16 states may thus encode four bits of information. Note that we are encoding such that the actual transmitted information rate is less than the final throughput: The baud rate is less than the data rate. The above-described system is often referred to as 16-QAM.

It is perhaps interesting to note that 4-QAM is just the same as QPSK and it may be demodulated as any other PM wave. 64-QAM and higher factors have been contemplated for and used in real systems. QAM with high encoding ratios may perhaps be best demodulated using Fourier-transform techniques.

Peak-to-Average Ratio of Multi-Tone Modulation

In many of the data communications formats above, quite a few significant frequency components are present and this poses a problem: Signal envelope may occasionally rise to high values while the average value remains relatively low. This so-called "peaking problem" rears its head in every peak-power-limited transmission medium, including radio transceivers and telephone systems. The peaking problem is a significant limitation to increasing data rates through these media.

Peak-to-average ratios are proportional to the number of tones used. A two-tone signal, for example, produces a peak-to-average power ratio of two or 3 dB. Higher numbers of tones produce higher ratios, reducing the power available in information-carrying components. Some schemes utilized in digital television and various military systems use special techniques to mitigate this effect. Refer to the References at the end of this book for more information.

Digital Coding Systems for Speech

Let it be clear at the outset that the goals of coding for error correction are in direct conflict with those for bandwidth reduction. They are discussed together here, though, because they embody many of the same principles, although those principles are used in different ways. Coding of speech waveforms for digital transmission is currently a hot topic of research and is providing a lot of motivation for further work. Let us first briefly review some basic digital methods for handling transmission errors.

Error Detection and Correction

The ability to detect errors in a data stream implies that information has been added to that data stream beyond what is required to convey the actual data. One of the first ways of doing this was the *parity-bit* system. A single bit is added to each byte that indicates, modulo-2, the number of ones the byte contains: its parity. Another way is to force all legal symbols to have a predefined parity, such as in AMTOR. No matter how you slice it, though, bits must be added to get error-detection capability.

Obviously, it is desirable to minimize added information and a variety of slick coding schemes have been devised over the years. For a few bits more, *error-correction* capability may be added. Digital transmission formats are discussed later that are inherently very error-tolerant; those tend to obviate the need for error detection and correction. But let us now look at some rudimentary error-correction schemes.

One early error correction scheme is *forward error correction (FEC)*. In the lowest form of FEC, two entire copies of the data are always sent. It is a simple matter to compare the copies and find differences: That is the error-detection part. Of course, it is possible for an error to repeat itself in both copies, but that may be made extremely unlikely. When errors are found, it is a matter of deciding whether one copy contains the correct data. Combined with a parity coding system, FEC gets two chances to receive data with the correct parity. FEC is, therefore, a second-order error-correcting system.

Another simple coding system for error correction uses the correlation properties of a signal to add check bits to the data stream: It is called *iterated coding*.[26] Refer to **Fig 6.1**. A chunk of the data stream is arranged in M columns of N bits. Modulo-two addition is applied to both the rows and columns to produce M+N check bits, which are appended to the data block. Carries from the additions are discarded. Iterated coding obviously requires some data-block synchronization so that the check bits may be distinguished from the data.

Now let us presume that an error occurs in the reception of a single bit of the total number received, MN. Errors will also appear in both corresponding column and row check bits, pointing to the erroneous bit. Its state is reversed and the error is corrected. No more than one error in the block may be corrected this way, although data bits may be instantly made suspect by the appearance of more than one erroneous bit in both the column and row checks. Iterated coding carries a fairly large overhead for small data blocks because M+N is similar in magnitude to MN. When the coding matrix is made square so that M=N, the penalty in additional bits is just 2M and this is not so bad for large values of M. For a data block of 10,000 bits, for example, the overhead amounts to only 2%.

Longer data blocks are more likely to contain errors, though, in direct proportion to their length. So while the coding overhead is shrinking with the square root of the block size, the likelihood of an error is increasing directly with the block size. Propositions involving iterated coding on long data blocks quickly become risky business, therefore.

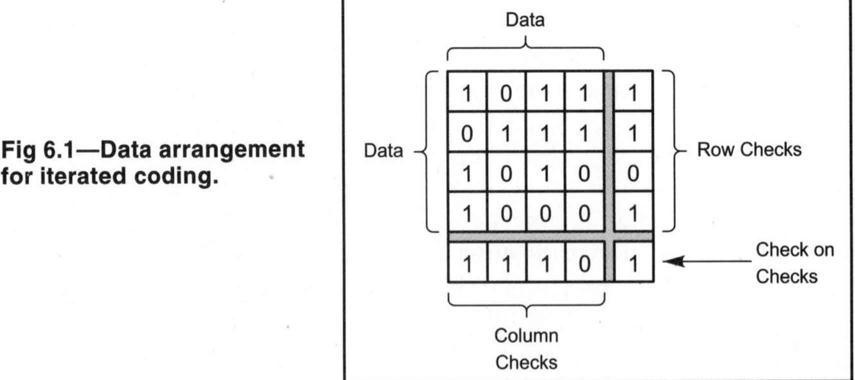

Fig 6.1—Data arrangement for iterated coding.

One term that appears in general discussions of error correction needs mentioning: *Hamming distance*. Hamming distance is the number of bits by which two binary numbers differ. For example, for the two numbers 00001111_2 and 00001000_2, the Hamming distance is three. This quantity may be related to the ability of codes to detect and correct errors in data by the following theorem: If the Hamming distance for two coded sequences is h, then it is possible to detect and correct as many as (h-1)/2 errors in one of the received sequences. Very many sophisticated codes have been discovered that are quite good at detecting and correcting multiple errors in data. These, like iterated coding, are *block codes* since they operate on discrete chunks of a data stream. Also possible are *linear codes* that operate on the fly in a bit-by-bit fashion. All these are well covered in the literature and will not be discussed further here.

Data Compression

In his proof of fundamental information theory, Shannon considered only the average probability of error over all codes of length n, chosen at random. He showed that error probability goes to zero as n goes to infinity. An as-yet-unsolved problem in information theory is to produce a code that satisfies Shannon's theorem and to mathematically prove its effectiveness. That has been the subject of a great deal of work.

Quite often, not all possible combinations of bits are used in the data, as is the case for words in English. We cannot just make up combinations of letters and call them language. Each English word may be considered a code and thus may be represented by a *code word*. If our lexicon includes, say, N=65,536 words, then each word may be represented by $\log_2 N = 16$ bits. English has a heck of a lot more words than that, but you get the idea. Now this representation achieves a great measure of *data compression* since a typical 5-letter word would usually require 35-40 bits.

The above example shows that speech at 150 words per minute may be digitally transmitted at a rate as low as (150 words/min) $\times$ (60 sec/min)$^{-1}$ $\times$ (16 bits/word) = 40 bits/sec. That can be sent in a very narrow bandwidth, but none of the speaker's vocal qualities are transmitted. You cannot tell who is speaking unless they tell you. We do not see such a scheme actually implemented anywhere because of the difficulties of automatically recognizing spoken words and because of complex problems reproducing the speech by machine with the proper inflection, in context.

Another way to compress data is to find redundancies in them. A long, contiguous string of binary zeros, for example, may be represented by a single binary number whose value is equal to the length of the string. Another set of binary numbers may be used to represent contiguous strings of ones. This kind of coding has proven useful in facsimile transmission where the blank or white parts of a page greatly outnumber the black parts (text or line graphics). Alternatively, differential coding may be employed to reduce the number of bits required. For multi-bit symbols, the difference between adjacent symbols is almost always a smaller value than that used by the symbols themselves.

A compression strategy devised by Huffman is notable in that it tends to require the smallest-possible number of bits to represent a block of data. As an example, let us presume we are coding a block of English text. The procedure is as follows. First, arrange the symbols (the words or letters) in the data in order of decreasing probability. Perhaps the most common English word is *the*, so it appears at the top of the list. Second, take the two entries at the bottom of the list–the least common ones–and combine them into a new symbol whose probability is set equal to the sum of the individual probabilities. Third, place this new entry in the list by ranking its probability and eliminate the two original symbols. Lastly, repeat the second and third steps until only a single symbol remains that has 100% probability. *Huffman coding* may thus minimize the number of bits used to represent a block of information to the limit of entropy theory iteratively.

Recent work in coding for compression and transmission has been spurred onward by telecommunications carriers, the recording and broadcasting industries, and by the Internet. Coding applications for speech and video have received the most attention. We shall make speech processing the focus of the remainder of this chapter.

Digital Coding of Speech

Digital coding of speech is certainly the topic of ongoing research and it cannot be covered completely here. Let us examine some basic systems for it, though, that lend themselves especially well to DSP implementations. Some are new and some are quite old. They may be divided into three general domains: time, frequency, and time-frequency. That last domain is any sort of combination of the first two.

Time-domain coders operate on a sampled speech signal primarily on a sample-by-sample basis and are therefore concerned only with representing its wave shape. Frequency-domain coders model the production of speech through convolution and transmit spectral information. For purposes of this chapter, so-called *parametric coders* are included in this category. Time-frequency coders employ some combination of time- and frequency-domain coding and thus encompass most of the hybrid techniques. Let us begin our discussion with time-domain coders.

Time-Domain Coding: μ-Law

The public telephone network went digital a long time ago. Engineers decided that eight-bit quantization was good enough for *toll quality* and that was the system adopted; however, dynamic range is only about 48 dB without further processing and that was deemed insufficient. To extend dynamic range, a method of logarithmic encoding is employed called *μ-law encoding*. In it,

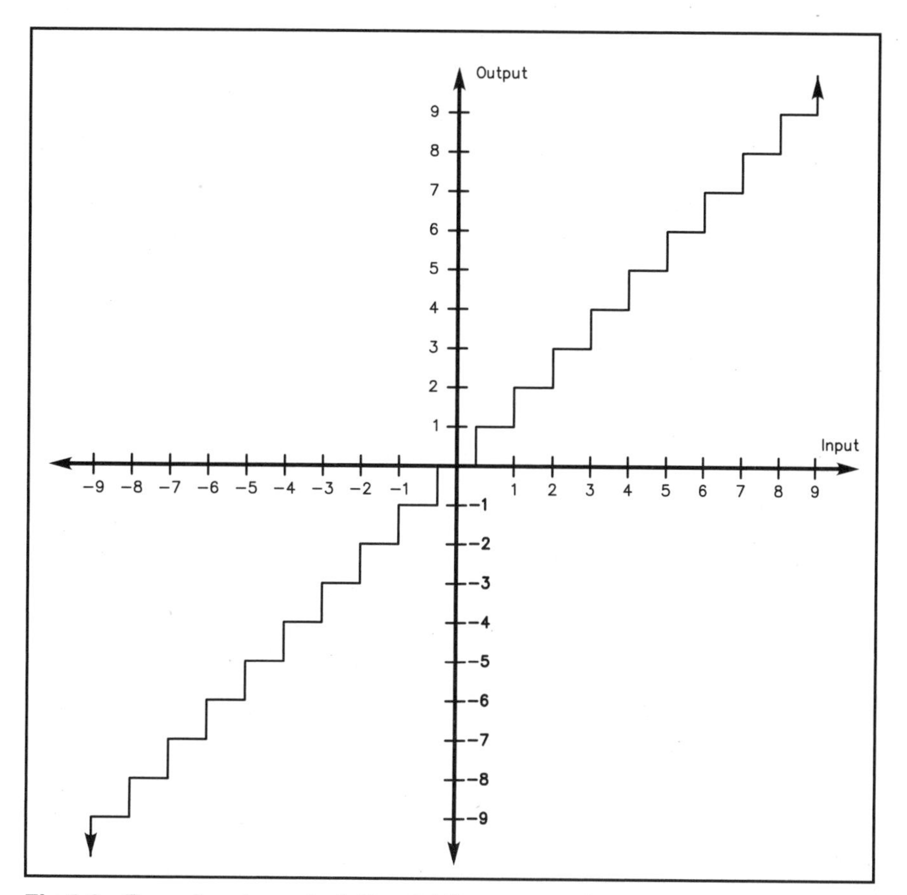

Fig 6.2—Transfer characteristic of a linear quantizer.

large amplitudes are quantized with coarse quantization steps, and small signals use smaller steps. This is an eminently reasonable way to get samples of an analog signal. It just makes sense; however, as one man said, "Common sense is not all that common." Workers on telephone systems dealt with all the complexities of network and information theory and came up with some remarkable discoveries that remain valid to this day.[27]

Instead of using 256 evenly spaced quantization steps, as shown in **Fig 6.2**, μ-law coding distributes the steps over the range so that a non-linear transfer characteristic is produced, as shown in **Fig 6.3**. The main drawback to coding this way is a lower limit on maximum SNR. To analyze SNR, we must first recognize that the equations presented in Chapter 2 for ADCs having a linear transfer function are no longer valid.

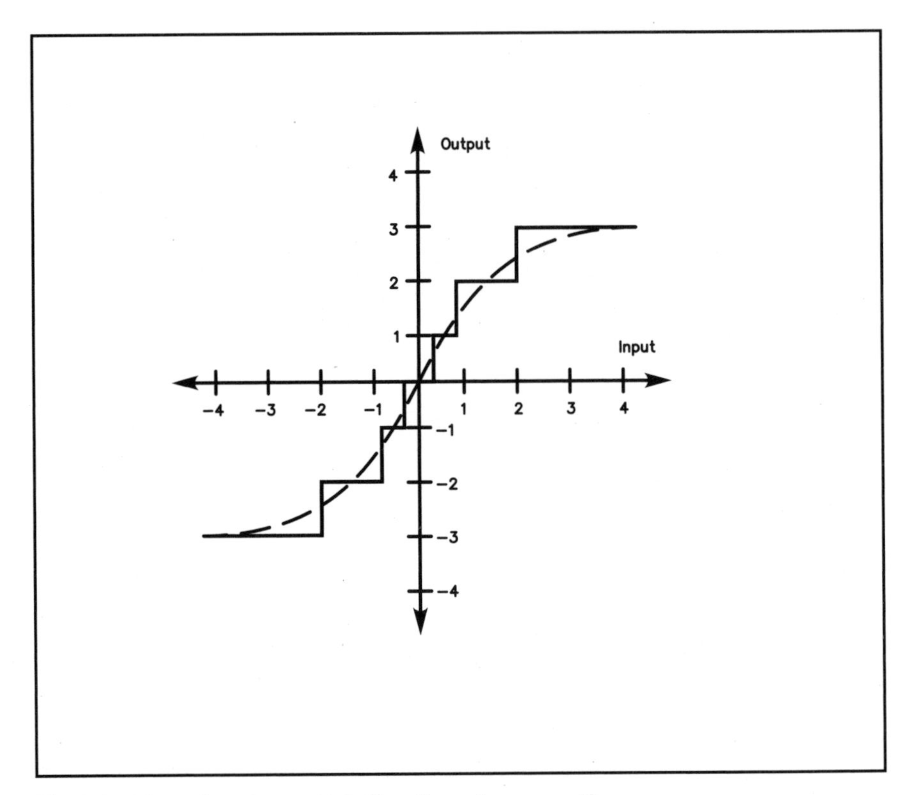

Fig 6.3—Transfer characteristic of a μ-law quantizer.

μ-law coding may be formulated by:

$$f(x) = \text{sign}(x)\left[\frac{\ln\left(1 + \mu\,|x|\right)}{\ln\left(1 + \mu\right)}\right] \tag{74}$$

where μ=255 for North America. Japan also uses this system. Some other countries use *A-law coding*, which is a little different. Dynamic range is found by solving Eq 74 for x.

Setting Eq 74 equal to the smallest-possible positive quantization step $1/2^{b-1}$ yields the smallest-possible input signal that can be resolved:

$$\frac{1}{2^{b-1}} = \frac{\ln\left(1 + \mu x\right)}{\ln\left(1 + \mu\right)}$$

$$x_{MIN} = \frac{\left(1 + \mu\right)^{\frac{1}{2^{b-1}}} - 1}{\mu} \tag{75}$$

This is much smaller than $1/2^{b-1}$ and for b=8 and μ=255, signals may be quantized at a very low level. Dynamic range is thus extended.

Derivation of an expression for SNR with this system is difficult but not intractable. As before, the noise voltage is the expected value of the quantization error over the distance between two adjacent quantization levels. One quantization level may correspond to an integer n divided by the number of possible levels, 2^{b-1}:

$$x_n = \frac{\left(1+\mu\right)^{\frac{n}{2^{b-1}}} - 1}{\mu} \tag{76}$$

and the next higher level occurs at $(n+1)/(2^{b-1})$:

$$x_n = \frac{\left(1+\mu\right)^{\frac{n+1}{2^{b-1}}} - 1}{\mu} \tag{77}$$

The difference between these two input levels is:

$$\begin{aligned}
\Delta x &= \frac{\left(1+\mu\right)^{\frac{n+1}{2^{b-1}}} - 1}{\mu} - \frac{\left(1+\mu\right)^{\frac{n}{2^{b-1}}} - 1}{\mu} \\
&= \frac{\left(1+\mu\right)^{\frac{n+1}{2^{b-1}}} - \left(1+\mu\right)^{\frac{n}{2^{b-1}}}}{\mu} \\
&= \left(1+\mu\right)^{\frac{n}{2^{b-1}}} \left[\frac{\left(1+\mu\right)^{\frac{n}{2^{b-1}}} - 1}{\mu}\right] \\
&= \left(1+\mu\right)^{\frac{n}{2^{b-1}}} x_{MIN}
\end{aligned} \tag{78}$$

and the expected value of the noise voltage is:

$$\sigma_n = \frac{\Delta x}{\sqrt{12}} = \frac{(1+\mu)^{\frac{n}{2^{b-1}}} x_{MIN}}{\sqrt{12}} \tag{79}$$

SNR is the ratio of the input level x_n to the noise:

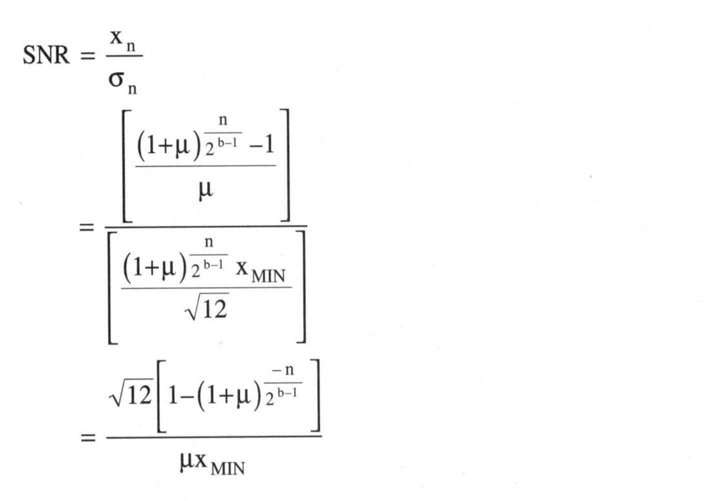

$$\begin{aligned} SNR &= \frac{x_n}{\sigma_n} \\ &= \frac{\left[\dfrac{(1+\mu)^{\frac{n}{2^{b-1}}} - 1}{\mu}\right]}{\left[\dfrac{(1+\mu)^{\frac{n}{2^{b-1}}} x_{MIN}}{\sqrt{12}}\right]} \\ &= \frac{\sqrt{12}\left[1-(1+\mu)^{\frac{-n}{2^{b-1}}}\right]}{\mu x_{MIN}} \end{aligned} \tag{80}$$

From **Fig 6.4**, a plot created using Eq 80, it is evident that this system does a fairly good job of holding SNR constant over the range of input levels as compared

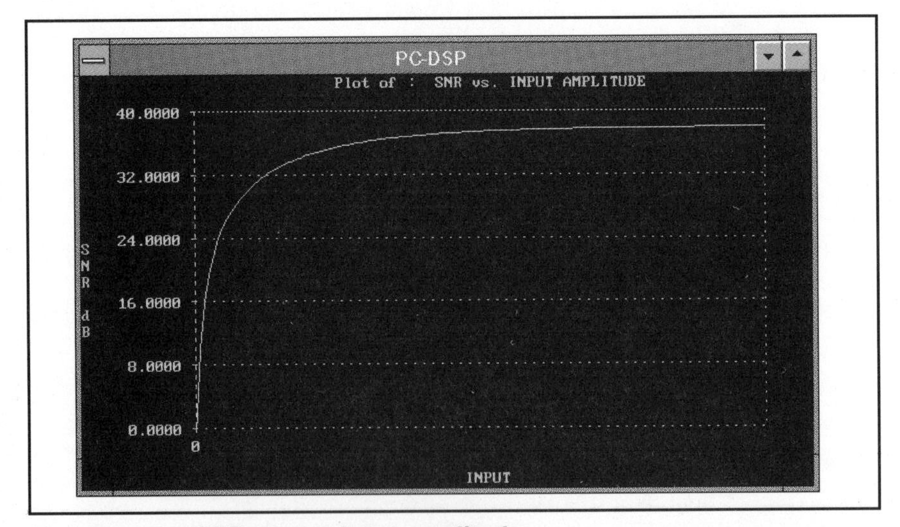

Fig 6.4—μ-law SNR-versus-input amplitude.

to straight 8-bit PCM. It is interesting that for μ=255 and b=8, input dynamic range is extended to about 75 dB, while output SNR cannot exceed 38 dB.

A typical T1 line connected to the public telephone network is designed to handle 24 simultaneous calls. At a sampling frequency of 8 ksamples/s, which is just sufficient to quantize 3 kHz of input bandwidth, an 8-bit quantizer produces 64 kbps. Now 24 calls need a serial bit rate of 24×64,000=1.536 Mbps. We may infer that a data connection using an analog modem at 48 kbps, occupying one of the 3-kHz channels, is not the most efficient use of resources, since were the digital interface extended to the user, 64 kbps would be possible. The conclusion is that the analog side of the public telephone system is not optimized for data traffic—no great shock, since that is not what it was designed for in the first place.

Some bright person came along and discovered that the bandwidth occupied by digitized audio could be reduced by coding the *difference* between samples rather than directly coding the samples themselves. Dynamic range required by the difference between samples is almost always less than that required by the samples themselves, and so may be coded with fewer bits, a lower sampling rate, or both.

Delta Modulation

First conceived after WW II, delta modulation (DM) reduces the number of bits used in quantization to one; that is, b =1. Sampling rates are typically close to those used on the public telephone network; however, it will be shown that toll quality may be obtained with certain DM schemes at much lower data rates.

DM encodes the difference between samples by employing the following algorithm.[28] When input voltage is increasing, the encoded bit is a one; when decreasing, it is a zero. See **Fig 6.5**. A fixed value of voltage change, dV, is associated with each bit. The encoded bit stream is integrated at the decoder to

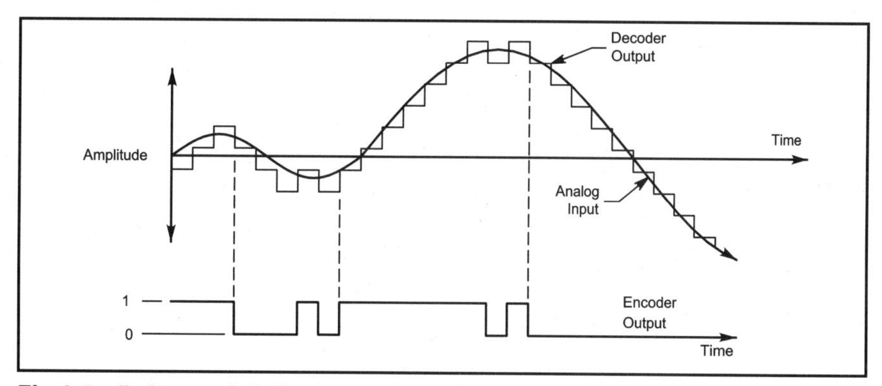

Fig 6.5—Delta modulation representation.

reproduce a copy of the original analog signal. Because dV is fixed, this scheme has a problem: It cannot reproduce signals having slopes that exceed the maximum integration time constant, dV/dt. This phenomenon is known as *slope overload.* Slope overload creates a roll-off in the high-frequency response of the system.

Increasing the value of dV helps mitigate this problem, but introduces *granular noise*. Large values of dV tend to mask small-amplitude signals. When input signals are smaller than dV, the quantizer output is an alternating sequence of ones and zeros and small signals are lost. DSP offers the possibility of adaptively changing the value of dV on the fly to address the issues of granular noise and slope overload.

Continuously Variable Slope Delta Modulation (CVSD)

CVSD, first introduced by Greefkes and Riemens in 1970,[29] adaptively alters dV based on the values of the last three or four bits in the stream. See **Fig 6.6**. When the last three or four bits are not identical, the system is equivalent to linear DM. When contiguous strings of ones or zeros occur of at least that length, though, the integration factor dV/dt is amplified to increase slope. Granularity problems are therefore no worse than in linear DM and slope overload is greatly ameliorated. Note that the figure incorporates an *adaptive predictor*, described in Chapter 8.

The adaptive character of CVSD results in a SNR curve similar to that produced by μ-law coding. A non-linear transfer function is created in this case because of the varying nature of dV/dt. I_1 is referred to as the *primary integrator* and I_2 as the *pitch integrator*. Time constants of 1 ms and 5 ms, respectively, have been found effective for speech. Considering its simplicity, the performance of this system is remarkably good. The quality of CVSD at 32 kbps is about as good as that of 64-kbps PCM.

We may analyze CVSD performance by comparing it with its cousin, linear DM.[30] When the input frequency-amplitude product does not exceed:

$$\left(A_{in} f_{in}\right)_{MAX} = \frac{f_s \, dV}{2\pi} \qquad (81)$$

the systems are the same. For single integration at the decoder, the quantization noise power is:

$$\sigma^2 = \frac{2dV^2 f_{BW}}{3f_s} \qquad (82)$$

where f_{BW} is the system bandwidth in hertz. For a signal at the slope overload threshold, signal power is maximized:

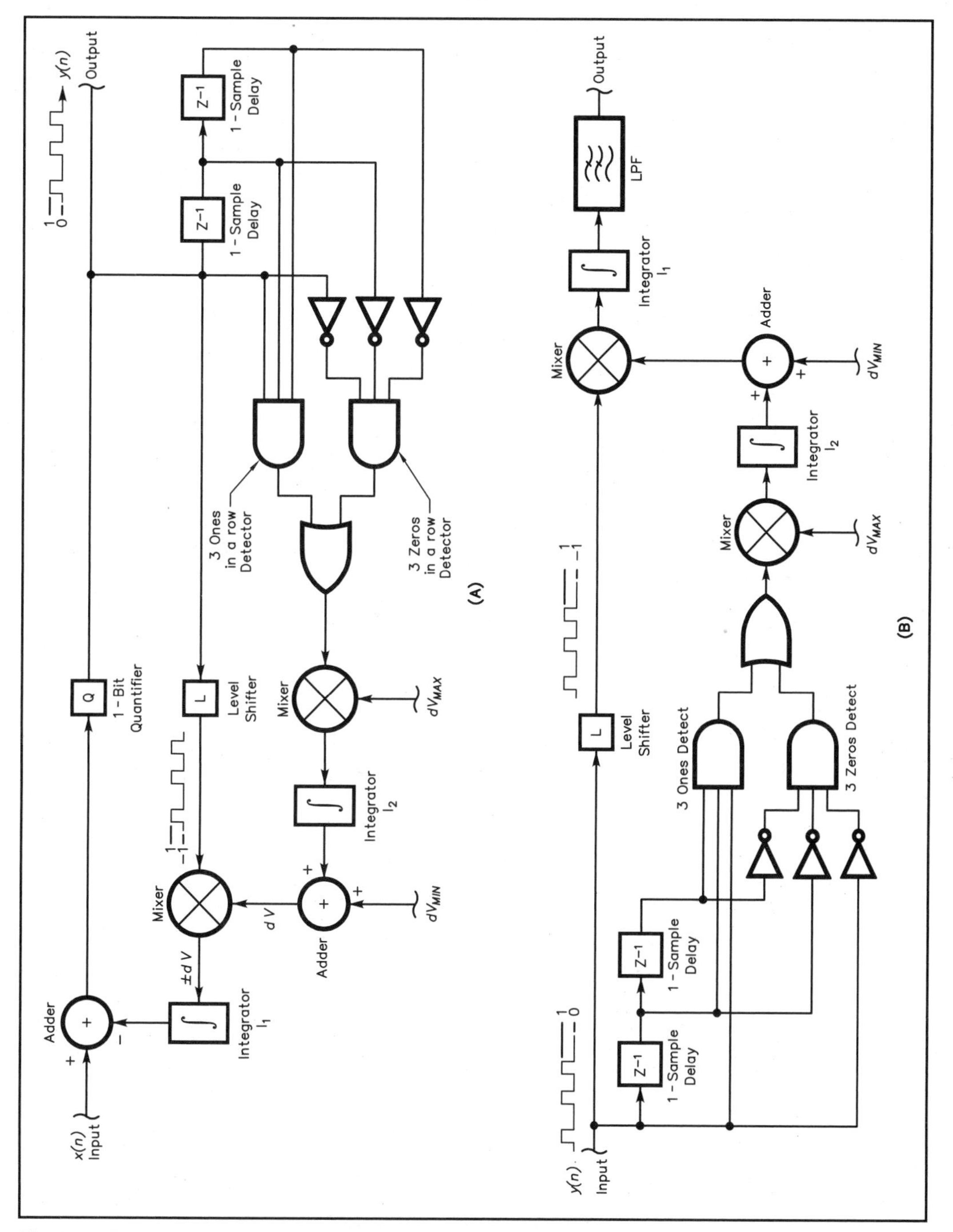

Fig 6.6—Block diagram of a CVSD encoder and decoder.

$$A^2_{MAX} = \frac{1}{2}\left(\frac{f_s dV}{2\pi f_{in}}\right)^2 \qquad (83)$$

and SNR is the ratio of Eqs 82 and 83:

$$\begin{aligned} SNR_{MAX} &= \frac{A^2_{MAX}}{\sigma^2} \\ &= \frac{3f_s^{\,3}}{16\pi^2 f_{BW} f_{in}^{\,2}} \\ &= 10\log\left(\frac{f_s^{\,3}}{f_{BW} f_{in}^{\,2}}\right) - 17.2\,\text{dB} \end{aligned} \qquad (84)$$

For f_s=32 kbps, f_{in}=1 kHz and f_{BW}=3 kHz, Eq 84 predicts an SNR of about 23 dB. It also predicts an SNR that varies in proportion to the cube of the sampling frequency and inversely with the square of the input frequency. Higher input frequencies produce lower SNRs. That effect is not very deleterious to speech signals because most of the energy in speech is below 1 kHz where SNRs between 30 and 40 dB are produced. Speech energy above 1 kHz is largely generated by plosive and fricative sounds like "p" and "f" that are atonal (noisy).

CVSD is quite tolerant of errors in the bit stream because any one-bit error simply pushes the waveform upward or downward slightly and does not affect its shape much overall. CVSD generally remains at toll quality up to bit error rates (BERs) of around 0.3%; μ-law coding requires BERs in the neighborhood of 0.01% for that quality. Also, CVSD may be operated at low bit rates with astonishingly good intelligibility. The analysis offered above shows that SNR degrades markedly at low bit rates; however, audio bandwidth may be maintained as the sampling frequency approaches the Nyquist rate. 8-bit, serial PCM cannot do that.

On increasingly congested communications networks, a major goal is to reduce digital data rates and therefore occupied bandwidth. A comparison of bandwidths employed above shows that digital speech tends to gobble up more bandwidth than the original analog signal. A 64-kbps signal, for example, occupies more than 32 kHz of bandwidth when it is reconstructed. That is more than 10 times the analog input bandwidth.

Coders described above operate on speech signals in the time domain. Notwithstanding the advantages of MSK and PSK modulation, much work has been done on digital coding of speech signals in ways that reduce the number of bits used while retaining desirable properties of the audio. Those include the ability to recognize who is speaking, their inflection, tonal qualities, and so forth. This has led to the invention of parametric and *predictive coders*, which generally operate on the spectral content of speech signals. Design usually begins with some analysis of the nature of the human speech-production system.

Frequency-Domain Coding

Frequency-domain coding endeavors to characterize human speech by transmitting information about its spectral content. That content is obviously changing with time; thus some overlap with time-frequency coders occurs. This section is primarily concerned, though, with those coders that model speech production as an excitation source coupled to a propagation medium. Specifically, the excitation is wind from the lungs that crosses the vocal chords and exits through the mouth and nose; the propagation medium involves the entire vocal tract and whatever external environment may be present between the talker and the listener.

Researchers have found over the years that describing this process in exact mathematical terms is extremely difficult. Too many variables present themselves to create a perfect model of speech production. Consider, for example, that most can discern one person's voice from another and even can tell when a particular speaker has a stuffy nose or other respiratory ailment. Apparently, a lot depends on how the shape of the vocal tract changes with time, temperature, nasal coupling and a myriad of other factors. Still, many frequency-domain coders take advantage of the fact that speech may be modeled as a sound source at the far end of a tube that has a time-varying shape and variable stops along its length.[31]

Fortunately, it is not necessary to solve all the equations that depict such a model to glean some understanding of what makes a particular person sound the way he or she does. Reasonable approximations and assumptions may be made regarding the properties of speech production that provide insight about how those properties may be exploited to advantage in coding.

Three basic types of human vocal sounds may be defined that simplify our discussion:

1. Air pressure from the lungs encounters some constriction in the vocal tract, producing a noisy sound. Those sounds are termed *fricative*. Examples: "f" and "h."

2. Pressure may be built behind a stop in the vocal tract, then released, to produce what is known as a *plosive* sound. Examples: "p" and "k."

3. Air from the lungs vibrates the vocal chords, producing vowel sounds or *voiced* energy. Note that this type of sound is usually what allows listeners to discern who is speaking and is what gives each person his unique vocal identity. The first two types of sounds may therefore be termed *unvoiced*.

That depiction of human speech production becomes rather complex when we consider that the excitation source, the propagation medium, and the nature of stops in the vocal tract may be rapidly changing with time. The model may be simplified by assuming that the changes are fairly slow—on the order of 20-40 ms in time. Further, certain assumptions about the nature of the excitation and propagation medium allow us to define and exploit their properties to achieve efficient coding.

A Model of Fricative and Plosive Sounds

Fricative and plosive sounds are noisy; a model for their production may use white or colored noise as the excitation source and a duration envelope may be used to complete their characterization. Certain fricative sounds are distinguished solely by their spectral signatures. The difference between "f" and "s", for example, is just the bandwidth and frequency placement of the noise source. Some rhyming words differ only by first letter "f" and "s." Some examples are "fill" and "sill," "fink" and "sink," "fell" and "sell." It is notable that those words normally would not be confused in English contexts. Consonants like "f" and "s" are not affected much by the human vocal tract because they are produced at or near the teeth and lips.

Other fricative sounds, like "h," are slightly affected by the vocal tract because they are formed farther within it. That consonant, though, does not sound much like any other and is therefore somewhat unique.

Many plosive sounds like "b" are formed at the lips and sound relatively soft. The consonant "p" is very similar to "b", and the difference is a subtle one, related to the exact duration of pressure release. Others, such as "k" and hard "g" are formed farther down the vocal tract and are characterized by a much more rapid release. The difference between those two is again subtle, and it relates to positioning of the tongue. The same is true of soft "g" and "j," which are indistinguishable.

Some consonants may be regarded as a combination of voiced and fricative sounds, such as "d," "v" and "z". Without voicing, those sounds would be indistinguishable from "t," "f" and "s." Fricative and plosive sounds may thus be characterized and sent in digital form by representing several types of colored noise and by some release envelope. This model is considerably simpler than the following model for voiced sounds.

A Model of Voiced Sounds

Voiced sounds may be characterized by a tonal excitation source and the shape of the vocal tract during production. The vocal chords are brought close together to create harmonic vibration as air from the lungs is forced over them. The shape of the vocal tract has as much to do with harmonic output as the nature of the vocal chords.

Some vowel sounds require vocal tract shape to be held constant, such as "e" in "feed" and "a" in "fall"; others require it to change, such as long "i" in "life" and long "u" in "vacuum". Sometimes, the change in shape is quite radical, such as in "now" or "royal." Consonants "r," "w" and "y" fall in the voiced category, since they do not involve fricative or plosive sounds.

Quite often, vowels at the beginning of a word require a mechanism akin to plosive sounds. A *glottal stop* is used, as at the beginning of "ink" or "apple". Like a *plosion*, voiced stops are characterized by a relatively rapid release of pressure.

Coding for Convolution

In frequency-domain or *convolutional coding*, excitation information is transmitted separately from information about the shape of the vocal tract. The excitation may be characterized by its spectrum and vocal-tract shape; either of those, in turn, by its impulse response or by its frequency response. The first problem is to separate the two from the original speech waveform. In so doing, those who study *phonetics* usually define certain terms that relate to spectral content and vocal-tract shape.

For voiced sounds, *pitch* is defined roughly as the frequency of the excitation source of a person's voice. This term shall be used differently in the study of human hearing below. For now, we shall consider it to be the lowest frequency at which the vocal chords are vibrating strongly. Various harmonics, both odd and even, are usually present in the source.

Pitch energy propagates through the vocal tract, and its spectrum is altered by the frequency response of the tract. Resonant frequencies of the tract are called *formants*. Formant frequencies change as the shape of the vocal tract changes.

Experimenters have found it useful to transmit speech signals using this model. Pitch, formant and noise frequencies may be extracted from the original voice signal through spectral analysis. Spectral analysis in DSP usually involves the FFT, discussed in Chapter 8. As the number of discrete pitches, resonances and noise bandwidths characterized can be reduced, the bit rate of a digital voice coder can correspondingly be reduced. A convolutional decoder reconstructs the speech wave by employing a model that is the inverse of the coder.

What the decoded result sounds like depends not only on the quality and quantity of information transmitted, but also on the nature of the human hearing system. Let us proceed to investigate what it is about human hearing that influences a listener's perception.

On the Nature of Human Hearing

Speech communication is crucial to our society. It conveys a sense of how someone feels, how they are thinking and some idea of who they are more than any other form. Nothing is more comforting than hearing the voice of a loved one in dire times. We may postulate, therefore, that this mode of telecommunications will never be supplanted.

Because of that suspicion, we can write that the secondary goal of any speech-coding scheme is to preserve those characteristics of speech that allow the listener to recognize the speaker, along with those nuances that are so important to the transmission of emotion. In other words, a coding system has to preserve certain distinctive qualities of speech so that we can't tell the speech was coded. What are those qualities and what is it about human hearing that influences perception?

In the study of the human hearing system, it must be clear that there is no objective means of measurement. All information about what someone hears–or doesn't hear–must be learned subjectively through the responses of the listener. All we can do is ask questions of a subject and attempt to infer something about the nature of sounds. Further, we have no guarantee that a particular stimulus will be perceived the same by one subject as by another. We must therefore define terms for measurement and perception differently and separately.

Sound *intensity* is a physical measure of the product of air pressure level and air-particle velocity.[32] Two persons equipped with identically calibrated instruments will measure the same intensity for any given sound. *Loudness* is the corresponding perceptual magnitude. It may be defined as "that attribute of auditory sensation in terms of which sounds can be ordered on a scale extending from quiet to loud."[33] One unit of loudness, the *sone*, is defined by subjectively measuring loudness ratios. A stimulus half as loud as a one-sone stimulus has a loudness of 0.5 sones. A 1-kHz tone at 40 dB sound-pressure level (SPL) is arbitrarily defined to have a loudness of one sone.

We are left to wonder how a unit based solely on individual perceptions can be useful, since so much variation exists from person to person. Furthermore, it has been found that the method of applying stimuli and of obtaining responses from listeners has a large effect on results. Loudness comparison of two equal-frequency tones, though, generally produces reliable and repeatable data. Loudness comparisons between dissimilar stimuli, such as between a pure tone and a polyphonic source, yield unpredictable results because of poorly understood subjective effects. It may be argued, then, that a quantification of loudness scaling (one sound is half as loud as another) is as good as absolute loudness matching (one sound the is same loudness as another). In addition, many researchers have observed that under certain conditions, binaural presentation of stimuli results in loudness doubling.[34] That is: Two equal-loudness sources—if they are far enough apart in frequency–are twice as loud as one alone. This rule has to be used with caution, though. Evidence exists that loudness addition is far from a perfect description of human hearing.[35]

Frequency is the measure of some sound's number of cycles per second; each of us may measure frequency identically using similar instruments. We may define *pitch* as the perceptual quantity corresponding to frequency. Pitch is to frequency as loudness is to intensity. Note that the relations among loudness/intensity and pitch/frequency are not necessarily linear, nor are the two perceptual measures independent of one another. Under certain conditions, the pitch of a constant-intensity sound may be shown to decrease with increasing intensity, even when its frequency is held constant.

As ably documented by Fletcher,[36] Stevens and Davis,[37] and others, loudness depends on both frequency and intensity. **Fig 6.7** shows some loudness contours. Each curve represents a constant-sone level. These data have been gathered countless times, but the basic revelation remains unchanged: The most

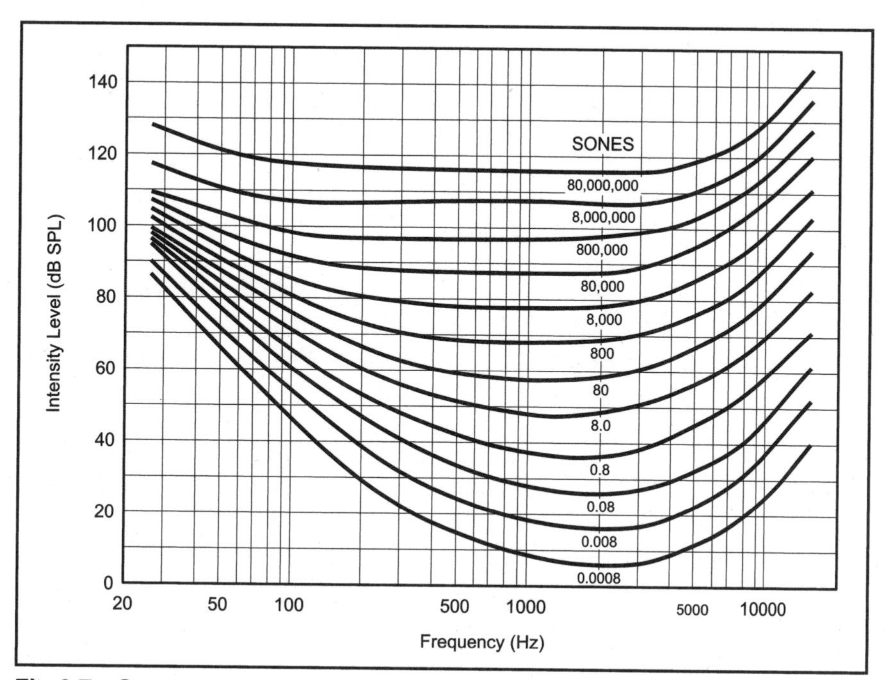

Fig 6.7—Constant-sone loudness contours.

sensitive frequency region of the human hearing system is between 1.5 and 3.0 kHz; and the curves get flatter as intensity is raised. Further, loudness grows faster with intensity at low frequencies. Finally, the curves reveal the dynamic range of human hearing: Tones below the zero-sone curve are inaudible, while tones above the top line are painful. In fact, we now know today that the useful dynamic range of human hearing is substantially less than shown. Extended exposure to sounds well under the top line may produce permanent hearing loss in some individuals.[38]

Let us now expand this discussion to include definitions for various perceptual thresholds, to introduce the idea of *masking*, and to present the concept of *critical bands*.

Thresholds of Hearing

One of the thresholds of hearing, the *intensity threshold*, may be defined as the lowest intensity a listener can detect. As shown before, though, we cannot directly measure the listener's perception; we can only ask questions of the listener as to whether he or she thinks the sound is audible. This might seem a fine distinction, but it turns out that the method of measurement determines the threshold as much as a listener's aural gifts.

At or near the intensity threshold, the subject's *criterion level* is in play. He or she might indicate some sound is audible when it *might* be present, or perhaps only when it is *definitely* present. With no incentive to produce correct results (such as large sums of cash), the criterion level may be beyond the experimenter's control.

One way of dealing with the criterion-level problem is to use a *criterion-free* experimental model.[39] The simplest of these is the "two-interval, forced-choice" paradigm. In this method, the stimulus is presented at random to the observer in one of two observation intervals. The subject is asked to determine in which of the two intervals the stimulus was present. A perfect observer always selects the interval that elicits the larger decision variable, thereby negating the criterion-level problem. He or she has a 50% chance of selecting the correct interval, even without actually detecting the stimulus. It may be shown that the *psychometric function* so produced solves the criterion-level problem.

We may also define *differential intensity threshold* as the ability to detect whether one sound is louder than another. Differential thresholds may be defined for other attributes of sounds, such as frequency and duration. A differential threshold is the amount some attribute must change to allow an observer to detect the change.

In the early 1800s, German physiologist E. H. Weber gave us a serious, quantitative depiction of differential thresholds. According to *Weber's Law*, differential intensity threshold is proportional to absolute stimulus intensity, I; or:

$$\frac{dI}{I} = k \qquad (85)$$

where k is known as the *Weber fraction*. This alleged constant also has been applied to sensitivity to changes in frequency and bandwidth, as well as non-auditory sensations such as color, image sharpness, pain, smell and taste. Very soon after Weber made it public, folks found out the rule broke down at intensities near absolute thresholds. They suggested a *modified Weber's Law*:

$$\frac{dI}{(I+I_0)} = k \qquad (86)$$

where I_0 is a constant. It's a good approximation, but apparently it doesn't hold exactly.

Masking

Masking is defined as the ability of one sound (the masker) to render another sound (the desired) imperceptible when present simultaneously or closely in time. It is quantified as the difference between the absolute intensity threshold of the desired in the absence of the masker and the elevated intensity threshold when the masker is present. Fletcher and Munson made a landmark study of the relation between loudness and masking effects.[40] They found that quiet sounds that are close in frequency to dominant sounds are rendered inaudible in proportion to their spectral separation and relative intensity. They were among the first to use bands of colored noise as maskers. An important effect is the relationship between the masker's bandwidth and the amount of masking. This relation is most prominent when the desired signal lies within the masker's bandwidth. Noise whose entire bandwidth does not overlap the desired signal does not contribute much to its masking.

This is one manifestation of the human hearing system: For many auditory functions, the ear-brain combination behaves as if it were a set of band-pass filters and energy detectors. Those filters are said to occupy *critical bands*.

Critical Bands and Peripheral Auditory Filters

One example of that behavior is provided by SSB over HF, wherein the ear quite often encounters severe phase distortion. The ear seems to tolerate reasonably large shifts in the relative phases of speech components without impairing intelligibility when the components are far enough apart in frequency. Scharf[41] defined the critical bandwidths associated with those theoretical auditory filters as "that bandwidth at which subjective responses rather abruptly change." He measured critical bands using two-tone masking techniques. Zwicker *et al.*[42] measured phase sensitivity using polyphonic sounds. These studies agree fairly well with others performed over the years. **Fig 6.8** is a plot of critical bandwidth versus frequency that averages the Scharf and Zwicker data.

Those and other studies support the idea that *differential frequency threshold* increases with frequency. In other words, it is more difficult to discern small frequency differences at high audio frequencies. Such a revelation means that it makes sense to use a frequency analysis system for speech whose frequency resolution matches that of the human hearing system. This sort of approach is eminently practical in DSP implementations.

Time-Frequency Analysis of Speech for Bandwidth Compression

Of course, speech coders must concern themselves not only with the total frequency content of a signal, but also how frequency content changes with time. Below, we shall explore aspects of speech signals in those two dimensions together.

Review of Traditional Spectral Analysis Methods for Speech

Choosing how to represent a signal is an important problem in DSP—just as important as how a signal is manipulated. The fast Fourier transform (FFT) has traditionally played a major role in speech communications research. Portions of speech such as voiced or fricative sounds, for example, can be modeled as the output of a linear system excited by a source either periodically or randomly varying with time. The output of such a system is simply the product of

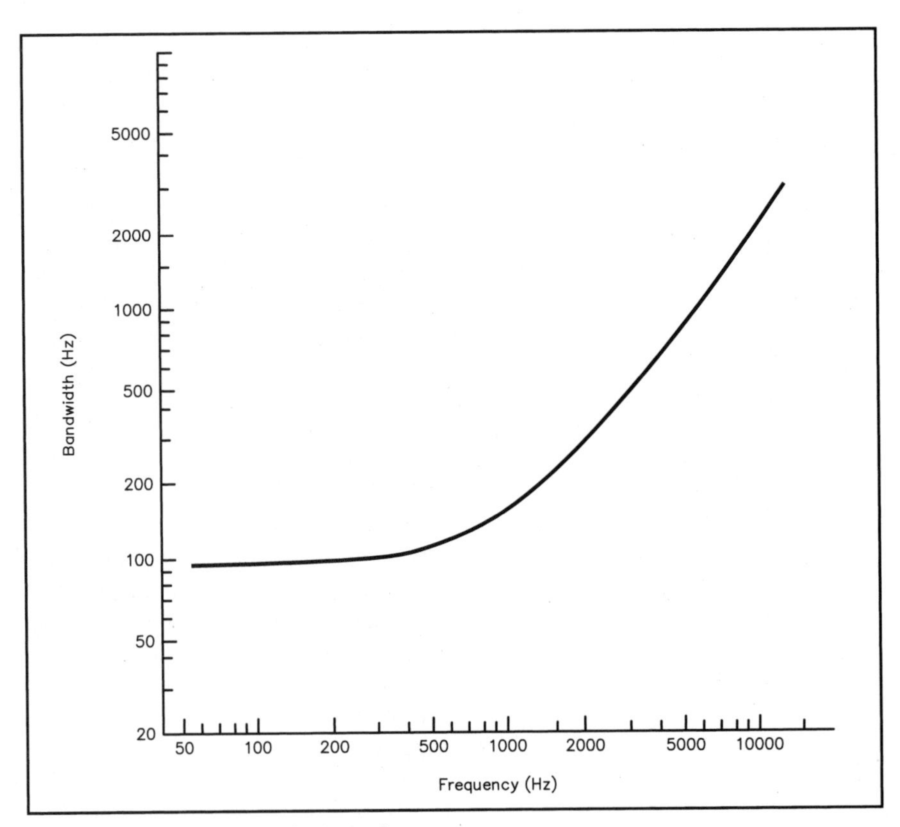

Fig 6.8—Critical bandwidth vs frequency.

the frequency response of the vocal tract and the spectrum of the excitation. Fourier analysis is useful in extracting these separate factors from speech waveforms, as noted above. Over the long term, though, speech signals are considerably more complex than this simple model. Thus, standard Fourier-transform representations that are satisfactory for periodic, stationary signals are not necessarily appropriate for speech signals whose properties rapidly and distinctly change with time.

As it turns out, it is reasonable and convenient to assume that the spectral content of speech doesn't change much over short time intervals, say 30 ms or so. This key unlocks a door to possibilities for bandwidth-compression of speech.

Details of FFTs are covered in Chapter 8. For now, let it suffice to write that the FFT is a *block transform*; that is, it operates on a block of input samples and produces a block of output samples that portray the frequency content of the input. Note that the FFT has fixed frequency resolution directly proportional to the sampling frequency and inversely proportional to the length of the input block. It is reasonable to suspect that an algorithm exploiting the fact that differential frequency threshold is somehow proportional to frequency would be more efficient than a straight FFT for the analysis of speech signals, because the total number of frequencies analyzed would be greatly reduced.

Critically Sampled Filter Banks

Digital signal processors are quite often optimized for the computation of convolution sums of the form:

$$y_n = \sum_{k=0}^{L-1} h_k \, x_{n-k} \tag{87}$$

Such calculations are the MAC calculations defined in Chapter 4 as those required to implement finite-impulse-response (FIR) filters. FFT algorithms don't necessarily make good use of MACs, so any filtering operation that reduces the complexity of subsequent FFTs is usually beneficial.

Rabiner and Schafer at Bell Labs worked on what are now called *multi-rate filter banks*.[43,44] In the first step of one such scheme, the signal under analysis first passes through two filters: a high-pass and a low-pass. See **Fig 6.9**. These filters have nearly identical cutoff frequencies and thus separate the input spectrum into high- and low-frequency bands. Since each filter's bandwidth is half the original signal's bandwidth, the sampling rate at each filter's output may be reduced by a factor of two without destroying information (see Chapter 4). Decimation filters with bandwidth equal to half the input bandwidth (one quarter the sampling frequency) are called *half-band* filters. When correctly designed, they have certain properties that lead to further computational savings.

In the second step, the decimated high-pass output is saved for later processing. The decimated low-pass output is further split into two sub-bands using half-band filters as before. The high-pass output is saved and the low-

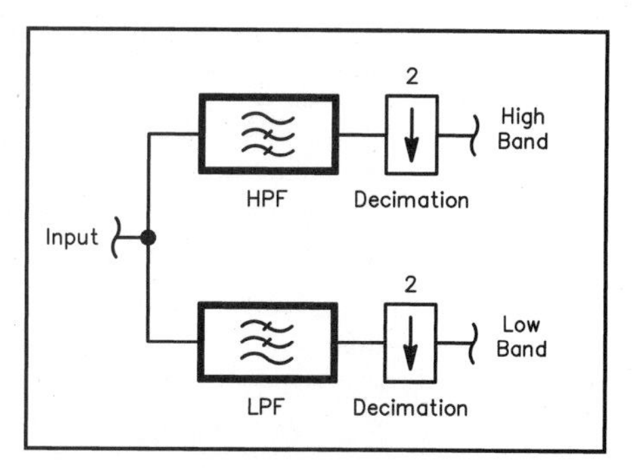

Fig 6.9—A band-splitting arrangement for multi-rate filter banks.

pass output split again. This process is continued until no further band splitting can occur. The result is shown as the block diagram of **Fig 6.10**. This is known as a *tree-structured* filter bank. The output of each filter is said to be *maximally decimated* or *critically sampled* because its sampling rate is minimized.

Note that the sampling frequency is halved at each step; hence, the number of samples available in any particular time span is also halved. Band splitting must end when we are left with only a single sample. Except for the final division, outputs from the system all come from the high-pass filters. These are further processed by FFTs that compute frequency content. This decomposition has made it easier to achieve good frequency resolution at the lower frequencies since fewer samples represent fewer frequency bins of an FFT applied there. Because the FFT is a block transform, the size of input blocks for each FFT (*i.e.*, the time span) is directly related to the size of blocks coming through the filters. Perhaps this is easier to fathom by studying the following example.

Refer to Fig 6.10. Let's say the system's raw sampling rate is 31,250 Hz. The input bandwidth is, therefore, half that or 15,625 Hz. In keeping with the premise that speech doesn't change much over time spans on the order of 30 ms, we'll take that as the length of the input block at the left-hand side of the diagram. To get the whole thing to work nicely, it would be good if the input block contained a number of samples equal to an integral power of two. By inspection, 2^{10} input samples looks like a good number, since:

$$\begin{aligned} \text{time span} &= \frac{\text{number of samples}}{\text{sampling frequency}} \\ &= \frac{N}{f_s} \\ &= \frac{2^{10}}{31{,}250} \\ &= 32.768 \text{ ms} \end{aligned} \tag{88}$$

At the output of the first stage of filters, the sampling rate is reduced to $f_s/2$ = 15,625 Hz; the number of samples in 32.768 ms is now 2^9. The input signal is split into two bands: 0-7812.5 Hz and 7812.5-15,625 Hz. At the next stage, the number of samples in the low-pass path is reduced to 2^8 and the signal is split into bands 0-3906.25 Hz and 3906.25-7812.5 Hz. This pattern shows that we are going to perform $\log_2 N=10$ iterations of band splitting before we get down to a single sample.

Now we have ten 32.768-ms blocks of samples to analyze, each with a different number of samples. Let's start with the highest-frequency block, which we call Band 9. Its bandwidth is 7812.5 Hz and its sampling frequency is twice that. Were we to apply an M-point FFT to these data, we'd have a frequency

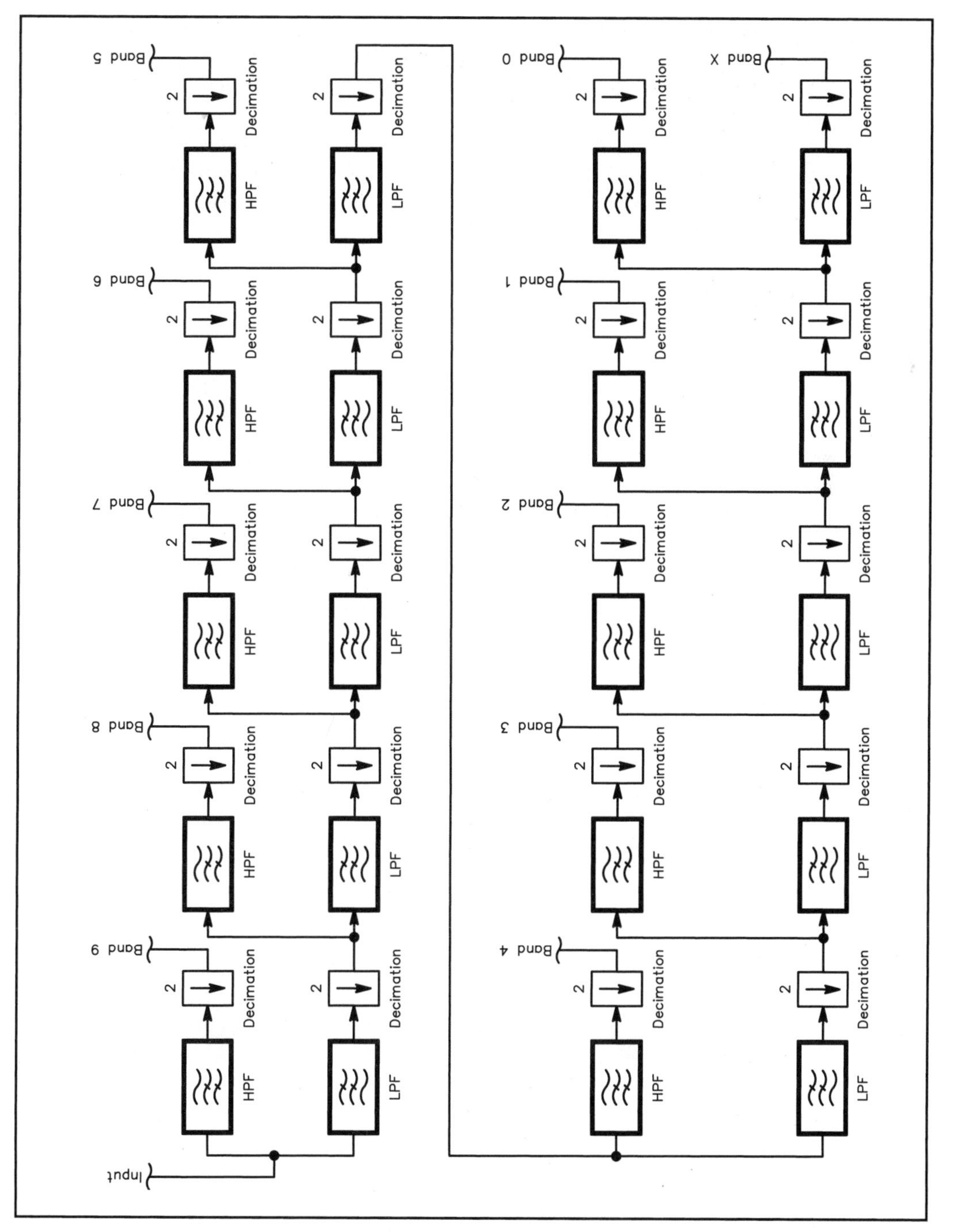

Fig 6.10—A tree-structured filter bank.

resolution of $f_s/2M$. Since the block length is N/2, we may perform N/(2M) FFTs on adjacent sub-blocks. In other words, we're confronted with a trade-off between good *temporal resolution* and frequency resolution. We're only interested, though, in the content of the entire 32.768-ms block: Its content doesn't change significantly during this period. From what we know about differential frequency threshold, we decide a frequency resolution of about 500 Hz is adequate for this sub-band. FFT size M therefore need only be:

$$\begin{aligned} M &= \frac{f_s}{\Delta f} \\ &= \frac{15{,}625}{200} \\ &= 32 \end{aligned} \tag{89}$$

For a real input signal, this produces 16 analysis frequencies or *bins*. Actually, 32 bins are produced, but the bins in the top group of 16 are just the complex conjugates of those in the bottom group, and so are redundant. Here we are with a block of $2^9 = 512$ samples and needing only 32 for our frequency analysis. Simple and direct would be to compute the 512/32 = 16 FFT blocks and average them. But as it turns out, we may select virtually any contiguous 32-sample block from within the input block, since frequency content doesn't change much over the input block.

So, for Band 9, a 32-point FFT was taken on the 32-sample block that was harvested. See **Fig 6.11**. We now know the frequency content of this band over a 32.768-ms period to a resolution of:

$$\begin{aligned} \Delta f &= \frac{f_s}{M} \\ &= \frac{15{,}625}{32} \\ &= 488 \text{ Hz} \end{aligned} \tag{90}$$

The same process is performed on Bands 6-8. At Band 5, no block harvesting is necessary since the decimated block is already 32 samples in length. When we get to Band 4, we run into a little snag: The block is only 16 samples long. It is tempting to just perform a 16-point FFT on this block, but then the frequency resolution would be $f_s/16 = 488/16$ 30.5 Hz, or the same as for Band 5. This is a getting a bit higher than the Weber fraction, so we decide to analyze this sub-band over a time period twice that of Band 5, or 65.536 ms. We then have our 32 samples and twice the frequency resolution.

Likewise with Band 3, doubling again requires a block-length increase to 131.072 ms to get the 32 samples and a frequency resolution of about 7.6 Hz. This band represents a frequency range of roughly 122-244 Hz— getting pretty

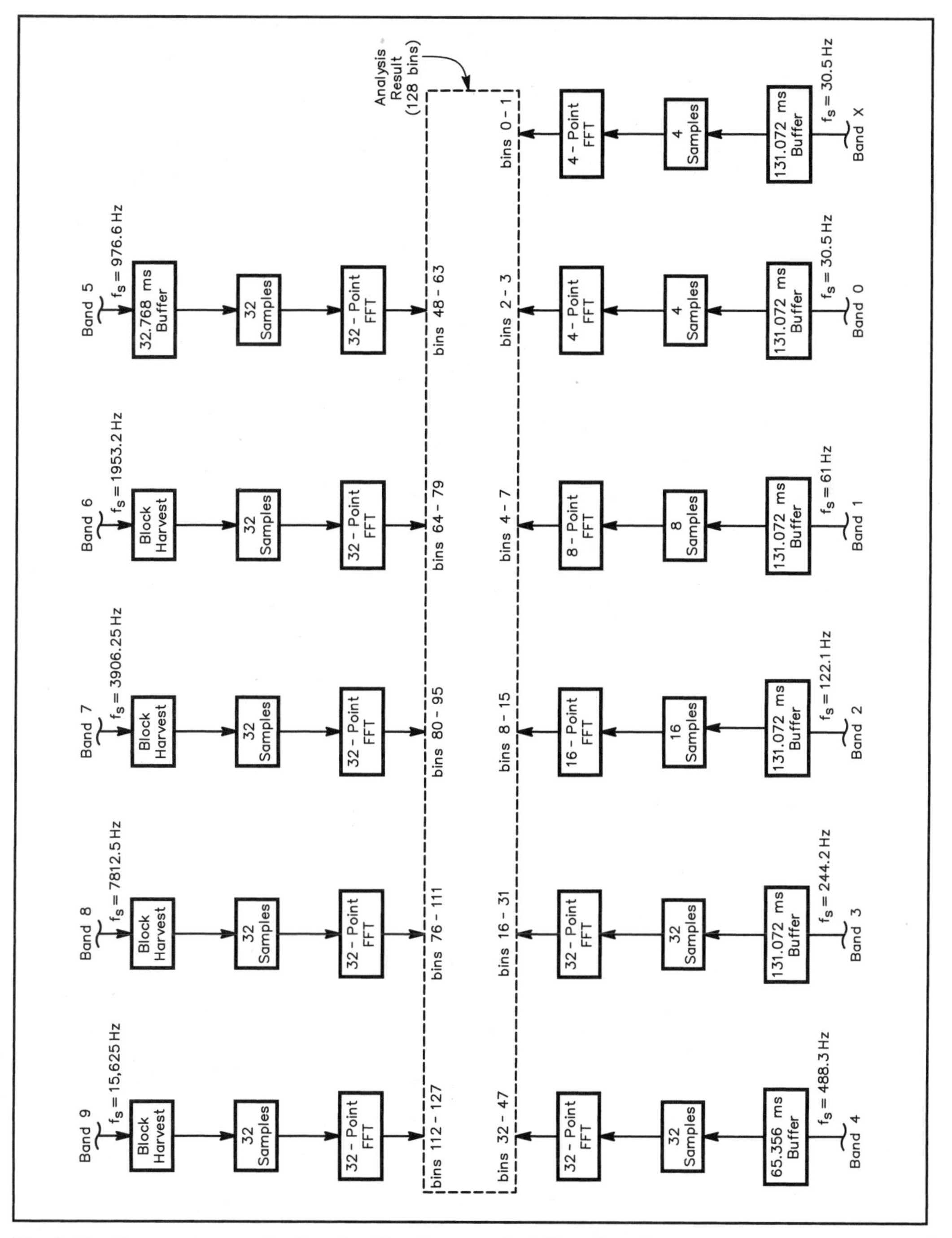

Fig 6.11—Frequency analysis of critically sampled filter bands.

low. For frequencies below 122 Hz (Bands 0-2), 7.6 Hz is deemed to be sufficient resolution and smaller FFTs are performed on 131.072-ms blocks. Band X, an 11th band, is just the leftover LPF output from the split that produces Band 0.

This alteration of sub-band block lengths reflects the main axiom under which the system operates: Good temporal resolution is more valuable than good frequency resolution at higher frequencies; at low frequencies, good frequency resolution is more important. This is supported by many of the studies cited previously and by common sense. One conclusion is that above a certain level, improvement in temporal resolution is useless because speech doesn't contain information changing so rapidly; further, the human hearing system cannot distinguish the rapid changes in spectral content that would be produced. Below a certain frequency threshold, improvement in frequency resolution is useless because the information contained in low frequencies is limited. The theory of natural selection[45] seems to indicate that animals do not develop their senses beyond what is necessary. It is therefore no surprise that our hearing matches our ability to communicate verbally. Animals in the wild are a somewhat different story, since they must be able to detect the presence of their enemies through subtle sounds, smells, and visual attributes; still, it is found that surviving species acquired the necessary tools and many of those that are extinct did not.

Perceptual Transform Coding (PTC)[46]

Perhaps some readers have experienced Internet audio systems, many of which use perceptual audio coders in one form or another. A data stream at 33.6 kbps occupies a bandwidth of at least 33.6/2 = 16.8 kHz (when reconstructed) and we know this can be coded in an analog format so as to fit through a 3-kHz-BW telephone line. This approximately 5.5:1 compression ratio shows that there is hope for significant speech compression.

As early as 35 years ago, attempts were made to reduce analog speech bandwidth by brute-force methods that squeezed all spectral components closer together in frequency.[47] At that time, the fast Fourier transform (FFT) was undergoing a rebirth. Perhaps it should have been evident from the nature of the beast that such frequency compression forces discrete transform elements to overlap, resulting in rather serious distortion. Perceptual coding cannot be obtained quite this simply. More-recent efforts utilizing sub-band coding and other methods[48] must have achieved at least some success, but we still don't see such schemes being employed generally.

Radio amateurs also have undertaken the quest for analog bandwidth compression. John Ash, KB7ONG, Fred Christiansen, KA6PNW, and Rob Frohne, KL7NA wrote about a system in *QEX* a few years ago.[49] Their premise was along the same lines as explained above: Certain parts of speech are redundant or irrelevant and so may be discarded. Tactics discussed below are a bit different from theirs in that spectral information is generally preserved across the frequency band of interest.

Anything reducing speech bandwidth by at least a factor of two ought to find immediate application in many services worldwide. It would reduce congestion on our crowded amateur bands as well as on commercial and military channels. It might increase telephone-circuit traffic manifold. It would not play in Peoria, though, unless it met the following quality goal: It must be difficult to tell the speech was coded. We must reluctantly infer, therefore, that previous bids have fallen short.

From the analysis above, we have samples in the frequency domain of a signal sampled in blocks 32.768 ms long. Now we shall construct an analog signal from those frequency-domain samples that is also 32.768 ms long but has a greatly reduced bandwidth. We will use the bins obtained from the sub-band decomposition above as the inputs to a 256-point inverse FFT (FFT^{-1}). An inverse FFT makes a time-domain signal (regular audio) from a frequency-domain signal. The sampling frequency of the output will therefore be:

$$\begin{aligned} f_s &= \frac{\text{number of samples}}{\text{time span}} \\ &= \frac{256}{0.032768} \\ &= 7812.5 \text{ Hz} \end{aligned} \qquad (91)$$

In so doing, the bins will represent frequencies spaced 1/0.032768 s = 30.5 Hz apart. The highest-frequency bin will correspond to the highest-frequency bin of the FFT done on Band 9. The next-highest-frequency bin will represent the second-highest-frequency bin of the FFT done on Band 9, and so on until all analysis bins have been *down-shifted* to their respective places in the coder's *synthesis* FFT^{-1}. Note that no temporal-resolution rules have been violated since each bin reflects the same block length in both FFTs. See **Fig 6.12**.

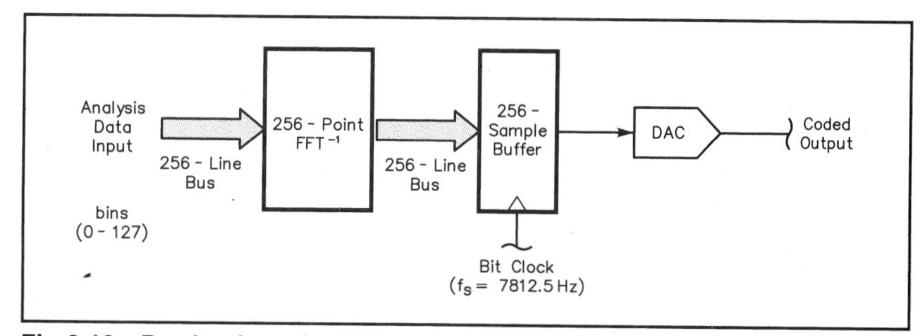

Fig 6.12—Production of a PTC-coded analog signal.

Frequencies of analysis bands are listed in **Table 6.1**; synthesis frequencies are listed in **Table 6.2**. Frequency resolution in synthesis is proportional to frequency. A speech signal of BW = 15.625 kHz has been coded into BW (30.5)(128) Hz = 3.90625 kHz. The frequency compression ratio is four. Note that this system, when restricted to half the input bandwidth, produces approximately the same compression ratio. An input bandwidth of 3.90625 kHz, for example, produces output bandwidth of about 977 Hz.

An additional, significant benefit of the system is that it may remove the restrictions placed on high- and low-frequency response by the characteristics of IF and AF filters in transceivers. Table 6.2's data reveal that extremely low frequencies are shifted upward by several hundred Hz. That means the low-frequency response of the system is preserved even when the coded signal passes

Table 6.1
Frequencies in PTC Coder Analysis

Band	*Low-Pass Range*	*High-Pass Range*	*Sampling Rate*	*Samples/ 32.768 ms*	*Freq Resolution*
9	0-7812.5 Hz	7812.5-15,625 Hz	15,625 Hz	512	488.3 Hz
8	0-3906.3	3906.3-7812.5	7812.5	256	244.2
7	0-1953.2	1953.2-3906.3	3906.3	128	122.1
6	0-976.6	976.6-1953.2	1953.2	64	61
5	0-488.3	488.3-976.6	976.6	32	30.5
4	0-244.2	244.2-488.3	488.3	16 (32 in 65 ms)	15.3
3	0-122.1	122.1-244.2	244.2	8 (32 in 131 ms)	7.6
2	0-61	61-122.1	122.1	4 (16 in 131 ms)	7.6
1	0-30.5	30.5-61	61	2 (8 in 131 ms)	7.6
0	0-15.3	15.3-30.5	30.5	1 (4 in 131 ms)	7.6

Table 6.2
Mapping of Frequencies in PTC Coder Synthesis

Band	*Input Freq Range*	*Output Freq Range*	*Number of Freqs*
9	7812.5-15,625 Hz	3418.0-3906.25 Hz	16
8	3906.3-7812.5	2929.7-3418.0	16
7	1953.2-3906.3	2441.4-2929.7	16
6	976.6-1953.2	1953.1-2441.4	16
5	488.3-976.6	1464.8-1953.1	16
4	244.2-488.3	976.6-1464.8	16
3	122.1-244.2	488.3-976.6	16
2	61-122.1	244.2-488.3	8
1	30.5-61	122.1-244.2	4
0	15.3-30.5	61.0-122.1	2
X	0-15.3	0-61.0	2
Totals:	0-15,625 Hz	0-3906.25 Hz	128

through two bandwidth-limiting filters: one in the transmitter and one in the receiver.

PTC Decoder

The decoder reconstructs the signal using exactly the reverse of the process used in the coder. See **Fig 6.13**. It first translates the signal to the frequency domain using a standard, 256-point FFT at a sampling rate of 7812.5 Hz. Input block length for the FFT is 256 samples or 32.768 ms. This produces analysis bins corresponding to 128 discrete frequencies. These samples are then inverse-Fourier transformed by band, with an additional provision for generating time-domain sequences longer than 32 samples for Bands 6-9 in that synthesis operation. The sequences are interpolated, filtered and combined in a manner opposite to that of the coder. The net result is a 32.768-ms block of output samples at the original sampling frequency of 31,250 Hz. Note that the bin order of each FFT^{-1} must be reversed; the subsequent interpolation and HPF operations (in Fig 6.13) invert the spectrum of the band being processed. The final output is obviously not a perfect reconstruction of the original input, since a compromise has been made between temporal and frequency resolution.

Computational Details

Let us look at some other processing details. Emphasis will be placed on computational efficiency. One heck of a lot of computation goes on in these algorithms. It is estimated that a PTC codec may be implemented on a dedicated DSP platform that has only modest processing power by today's standards. Without the shortcuts outlined below, much more horsepower would be required. Alternatively, increased processing capability would allow greater frequency resolution and therefore improved quality.

In the coder, the output of one filtering stage forms the input to the next. Enough output samples from one stage must be accumulated before the next stage's output can stationarily be computed. Further, the input buffer for a particular filter stage must grow beyond 32.768 ms by the length of the filter's impulse response. Finally, the filter's impulse response must be long enough to achieve *orthogonality* between sub-bands. This term means that no frequency component appearing in either the high-pass or low-pass sub-bands appears at significant amplitude in the other filter's output. That is, the filters must be sharp enough not to let frequency components appear simultaneously in both the high-pass and low-pass outputs. This requirement obviously presents itself most critically in and near the transition regions of the filters' frequency responses. Either some overlap or some exclusion of analysis frequencies must be tolerated, since short filters are not very sharp-skirted.

FIR half-band filters may be designed in DSP with impulse responses having odd-numbered coefficients equal to zero. See **Fig 6.14.** This is achieved using Fourier design methods when the total number of taps is odd.[50] The

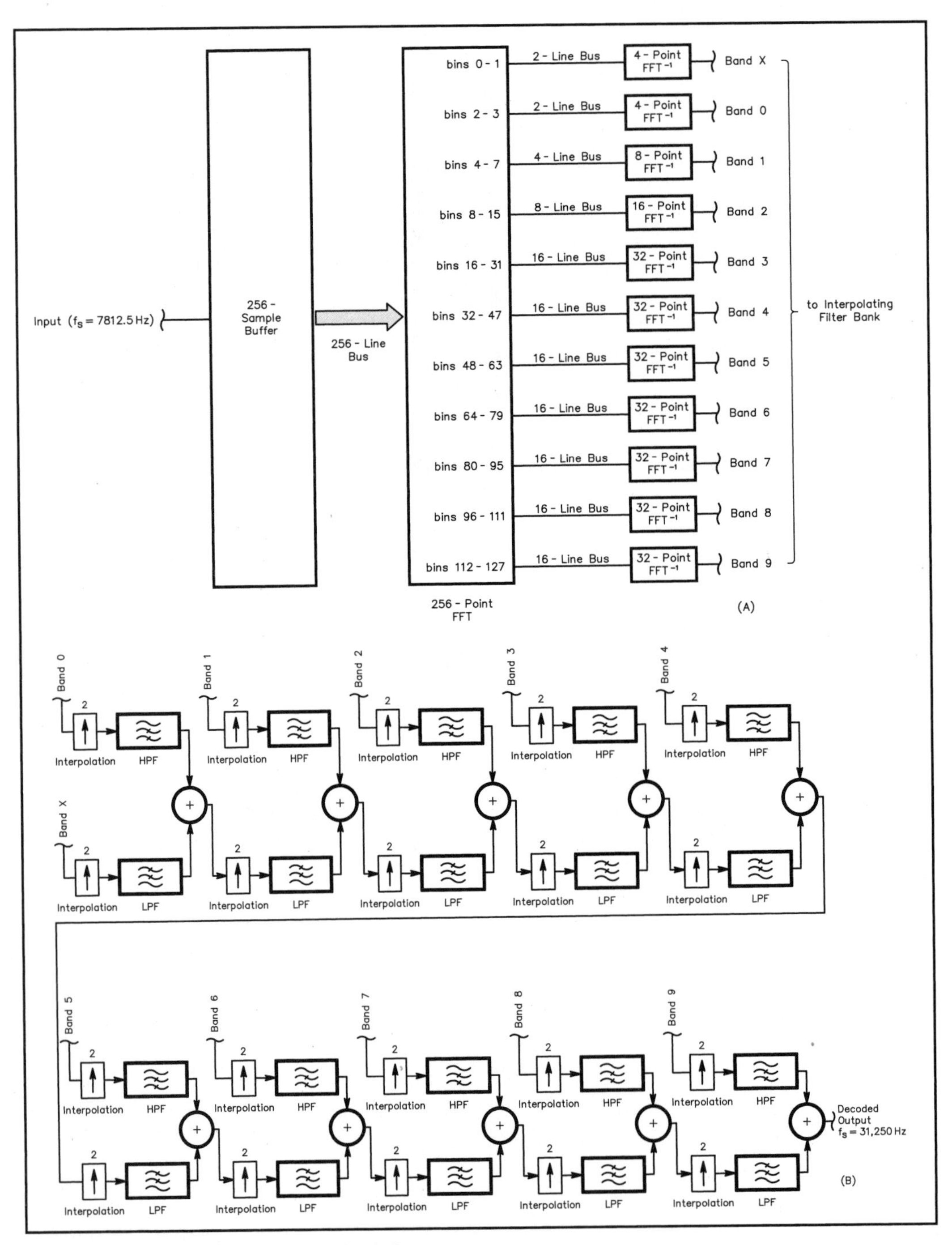

Fig 6.13—A complete 10-band decoder.

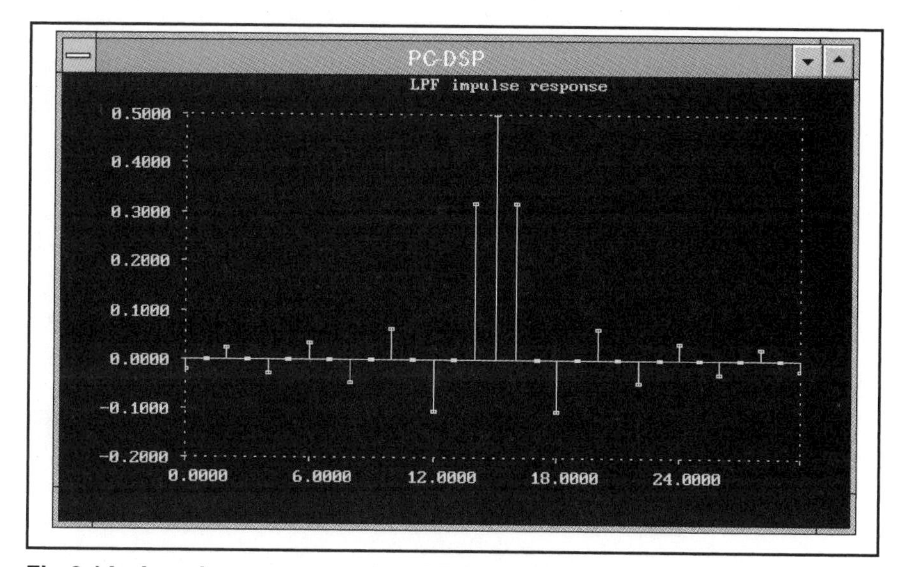

Fig 6.14—Impulse response of a LPF for L=33.

significance of that is that the total computational burden is reduced by a factor of two, since those taps with coefficients equal to zero don't need to be computed or added to the convolution sum. Also, it turns out that half-band, high-pass and low-pass filters may be designed so that their impulse responses are nearly the opposites of one another. For a filter of length L, the coefficients of a half-band high-pass filter, h_k, are simply the negative of the coefficients of a half-band low-pass filter, except for the coefficient at the center of the filter, $h_{(L-1)/2}$. See **Fig 6.15**. This reduces computational complexity by an additional factor of two, since the output of either filter is just the convolution sum using coefficients $\pm h_k$ plus the term produced using the center coefficient alone.

Filters of length L = 33 designed using a rectangular window barely meet the requirements above. To avoid having to insert delays in the FFT paths, it is well to store all the input samples for both coder and decoder before calculating all the filter outputs. This isn't always possible, though, since it results in significant throughput delay. Note that a small delay is always precipitated by the wait for buffers to fill. A complete filter stage using this *polyphase* approach is shown in **Fig 6.16.** The frequency responses of such typical FIR filters are shown in **Fig 6.17.**

Results

The coder and decoder are not synchronized. Because the 32.768-ms frames in the coder are not likely to be aligned in time with those in the decoder, a spectral-smearing effect always occurs. The magnitude of the effect depends

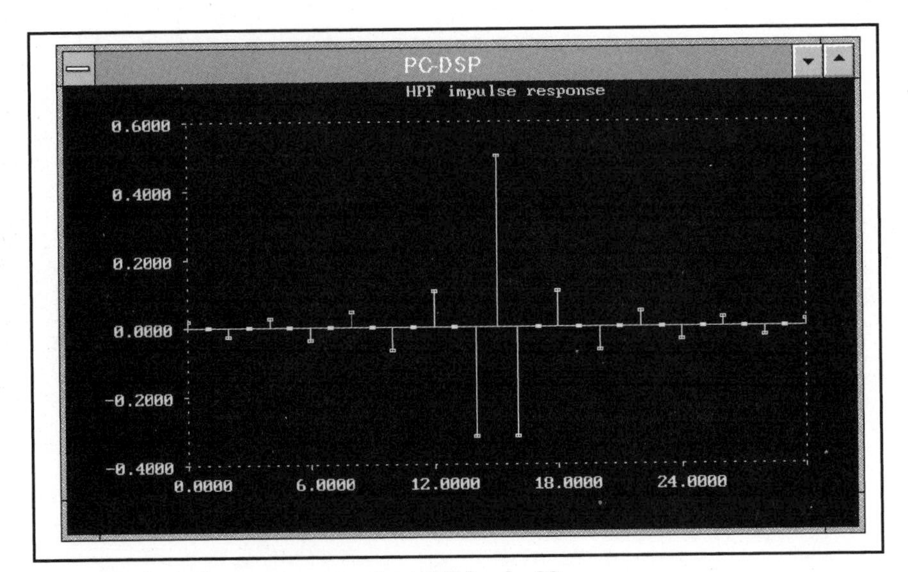

Fig 6.15—Impulse response of a HPF for L=33.

heavily on how much the high-frequency content of adjacent frames changes. As stated above, low-frequency content is not liable to change very much from frame to frame. In the worst imaginable case, high-frequency content changes markedly between frames and half the energy appears in one frame, the other half in the next. Total energy content is preserved, but the temporal resolution is compromised to the tune of half the analysis-block length. This effect has not presented itself as a perceptual problem during testing.

Originally, it was believed that PTC-coded speech compressed to one fourth its original bandwidth would still be intelligible, but it is not. The main reason

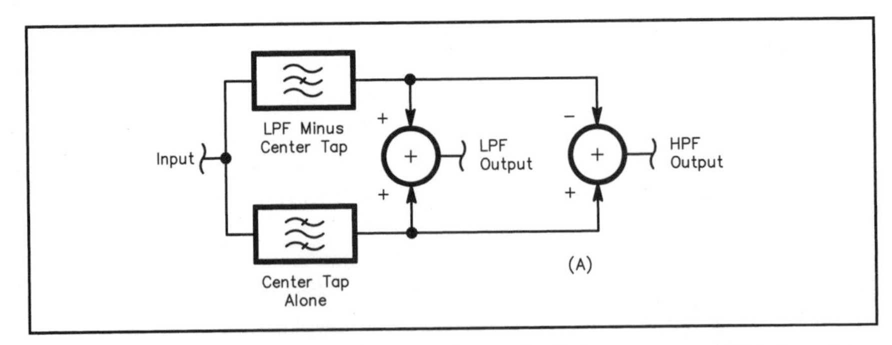

Fig 6.16—A polyphase filter that produces both low-pass and high-pass outputs.

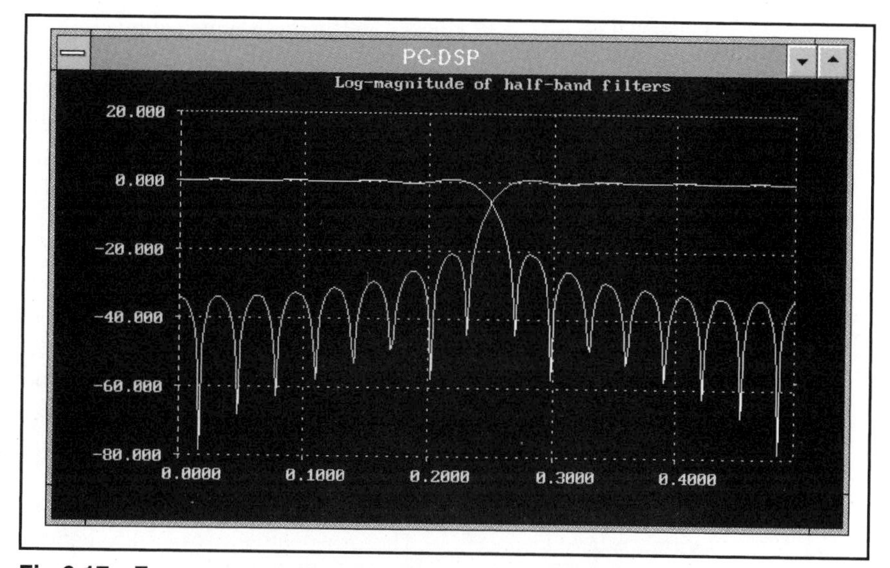

Fig 6.17—Frequency responses of the LPF and HPF.

for that seems to be that frequencies corresponding to the pitch of a person's voice are shifted upward in frequency too much to allow the ear to discern them. Formant energy resides much closer to pitch energy, rendering them indistinguishable from one another. That is not to say you can't still tell it's speech; it just sounds—well, different. Application of windowing to time-domain data in analysis and synthesis is the subject of ongoing experimentation. After years of listening to 2.4-kHz audio, it is astonishing how much the addition of some sibilance and presence improves perceived speech quality.

Many hams enjoy listening to SSB signals by using a much greater RX bandwidth than that used in the transmitter. Attribute this to the IMD products appearing beyond the transmitter's bandwidth that pass for sibilance at the receiver. Good thing they can't listen to the IMD products on the other side because the results would not be quite so pleasing.

You may say someone sounds like FM, but the trouble has been that the high-pass filters necessary to eliminate CTCSS tones have had a very deleterious effect on voice signals. More often, we are referring to the degree of quieting that is apparent. In the finish, PTC has some effect on signal-to-noise ratio as well.

Not only have we reduced bandwidth by a factor of four, but we have gained a signal-to-noise ratio (SNR) advantage of:

$$\Delta SNR = 10 \log 4 \approx 6 \text{ dB} \qquad (92)$$

Note that we've also avoided approximately 6 dB of QRM in the process (using appropriate IF filters) and that we have relieved our neighbors in frequency by the same margin. These factors apply to the on-the-air signal, not to the final result. Statistical noise from signal processing algorithms usually offsets the reduction in atmospheric noise. The system is subject to a magnified effect from any on-channel interference, if it is polyphonic. That is to say: If polyphonic, on-channel interference occupies bandwidth m, a PTC decoder will demodulate it with bandwidth = 4m. Selective-fading effects also are amplified by the same amount. PTC-coded speech is also a bit more susceptible to frequency errors.

As noted, the ear seems to be sensitive to the relative phase of components lying within the same critical band. This seems to be because such components may produce a beat frequency of greater than the critical bandwidth, resulting in an audible effect. It is interesting to hear how audio waveforms having different phase relationships between their spectral components—and that look quite different on a 'scope—sound remarkably the same.

In the form of an external audio processor, PTC would be compatible with virtually any transceiver. The prospective uses of the bandwidth savings are alluring, to say the least.

PTC codecs allow four or more times as many voice signals to occupy a given band as compared with non-coded signals. While this may not destroy all QRM, it sure seems to offer a better chance for radio operators to happily coexist. Application of PTC to other services, such as FM land mobile, is not quite so simple. Transceivers usually have synthesizer tuning steps of 12.5 kHz or 25 kHz to match the channel spacing and IF filters. New or heavily modified designs would have to be fielded to take advantage of greater spectral occupancy. Other uses may be made of the saved spectrum without changing channel spacing. Full-frequency-range stereo or four-channel speech, for example, is possible in typical voice bandwidths. With two or more independent channels, more information can be communicated. PTC may legally be employed now in many commercial services.

Since the vast public telephone network has already gone digital, some doubt arises as to whether it is useful there to increase traffic-handling capacity. Certainly, PTC coding could be applied prior to digitization to achieve a boost; however, large-scale rearrangement of multiplexing equipment would be necessary and lots of new gear would have to be purchased. In addition, it may be that speech-compression coding in digital form (after digitization) would be more cost-effective.

Summary

From the foregoing, it is evident that not all components of human speech are necessary to achieve high perceptual quality. Irrelevant components are sometimes made inaudible by masking or critical-band effects and therefore can be eliminated. Many modern digital coding methods make extensive use of these factors to achieve their efficiencies. We also find that speech does not change much from one short time frame to the next, and so contains redundancies. This is also used to reduce bandwidth.

It was shown that if the ear is less sensitive to differences in frequency as frequency increases, then the high-frequency territory is prime ground for bandwidth compression. The underlying principle of PTC is to analyze the frequency content of the input signal with non-uniform frequency resolution, to combine adjacent frequency bins that are closer together than the differential frequency threshold, and to down-shift some of the bins in frequency to create an analog signal of lesser bandwidth and frequency resolution. The processing power required to implement PTC is moderate by today's standards.

Direct Digital Synthesis

In the last two decades of the 20th century, *direct digital synthesis* or *DDS* created a revolution in the design of radio transceivers. The speed and noise performance of DDS make an attractive combination in many applications. DDS also has simplicity and inherent stability going for it. The term direct means that no feedback is involved. A PLL is an example of an indirect synthesizer. DDS is now found in virtually every modern rig and also in television sets, test equipment, cellular telephones and a variety of other products.

Basic Digital Synthesis

DDS is just the process of generating digital samples of an output wave, such as a sinusoid. In its most common form, a DDS is understood to also have a DAC to convert the sampled signal to the analog domain; but DDS is often applied where no DAC is involved. One example is during the generation of a digital BFO, as discussed below. Where DDS is used in analog frequency synthesis, though, a DAC is needed and its effect on DDS performance is included in the discussion of this chapter.

To generate a sinusoid digitally, it is convenient to use the relationship between its phase and its amplitude. Phase is directly proportional to time, so a phase accumulator (counter) may be incremented by a constant at each sample time to keep track of it. The value of that constant is directly proportional to frequency. Higher frequencies use higher increments at each sample time.

Each phase angle of all those possible corresponds to a discrete amplitude, according to the relation:

$$y_t = \sin\phi_t \qquad (93)$$

Now it is a reasonably simple matter to build a look-up table that uses the phase, ϕ_t, to index a digital amplitude value at each sample time. Those data may be held in a read-only memory (ROM) since they never change. With the appropriate sample clock and DAC added, this arrangement is shown in **Fig 7.1**.

The output of such a circuit is a step-wise representation of the desired sinusoid, as illustrated in **Fig 7.2**. From Chapter 2, we know that means the output is contaminated by aliases. They repeat at intervals of the sampling frequency and are normally removed using an analog LPF. That anti-aliasing filter would be tacked onto the right-hand side of Fig 7.1. Before discussing performance limits of DDS that are set by DACs and filters, a brief look at digital effects that limit performance is in order.

DDS Performance: Digital Effects

Since a DDS's phase accumulator has fixed numerical resolution, slight inaccuracies in the phase value must be tolerated. This is the same quantization problem that was covered in Chapter 2. A DDS suffers from phase distortion

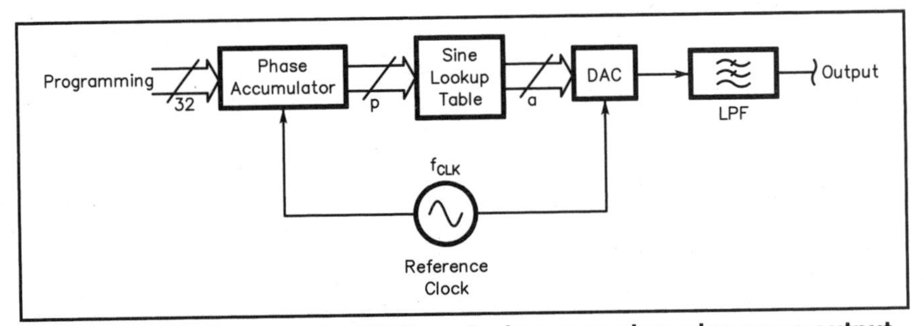

Fig 7.1—Block diagram of a DDS producing an analog, sine-wave output signal.

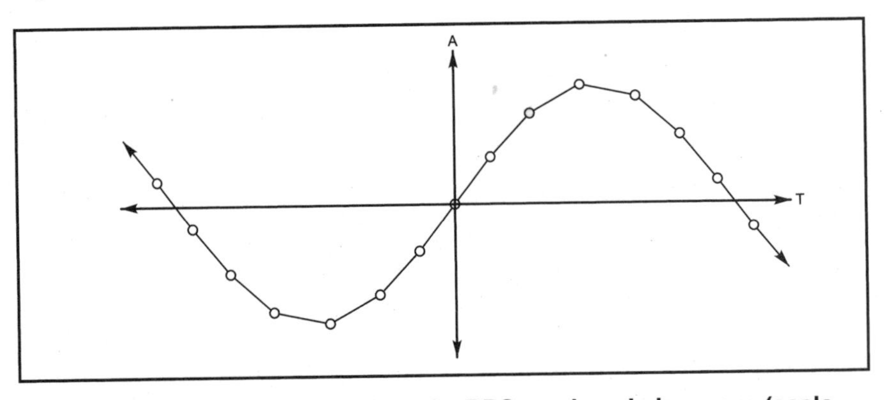

Fig 7.2—Step-wise representation of a DDS-produced sine wave (scale exaggerated).

that appears as quantization noise in the address of the look-up table. That distortion translates to undesired phase modulation on its output: phase noise. This phase noise is not quite the same as that produced at the output of a PLL circuit, though, since it is quantized and is liable to change markedly from sample to sample. Its pattern often repeats in a cyclical way, resulting in discrete phase-modulated (PM) spurs.

The output of a look-up table also has fixed resolution, so it is an imperfect representation of the correct output amplitude (most of the time). Quantization noise appears here in the form of AM spurs that again may concentrate themselves at discrete frequencies because of pattern repetition. The locations of these spurs is a complex function of the sampling rate, bit-resolution and output frequency. Their maximum amplitudes may be approximated, though, through application of sampling and modulation theories.

According to Cercas et al.,[51] maximum PM spurs may rise to:

$$P_{PM} = -(6.02p-5.17) \text{ dBc} \quad (94)$$

where p is the number of binary bits used to represent phase. AM spur levels may be as high as:

$$P_{AM} = -(6.02a+1.75) \text{ dBc} \quad (95)$$

where a is the number of bits used at the output of the look-up table. Note that phase errors may be slightly attenuated by using more bits in the actual phase accumulator than in the address to the look-up table. A 32-bit phase accumulator and 20-bit address are common. Also, a larger phase accumulator yields finer frequency resolution. Address resolution of greater than 20 bits produces very large look-up tables that are hard to fit onto even modern chips. A 20-bit-address table has 2^{20} or about 1 million entries; a multi-megabyte memory is thus required. The process of interpolation can achieve a large reduction in the size of look-up tables at the expense of a little more distortion.

The frequency resolution of a DDS is limited by the resolution of its phase accumulator, notwithstanding fractional-N techniques, treated briefly later. Frequency resolution of a DDS is:

$$\Delta f=f_{clk} (2^{-p}) \quad (96)$$

In this case, p is the number of bits in the phase accumulator, not the number of bits in the address. This resolution is constant throughout the frequency range of any particular device. That number df can be quite small: At a clock frequency of 10 MHz, a 32-bit phase accumulator gives a frequency resolution of 2.3 millihertz!

For analog synthesis operations, the appearance of amplitude data at the DAC must be synchronized so that all bits change at the same time. That is usually a function of the DAC itself, which is clocked at the output sample rate. Some researchers have found that differences in the time of switching of bits can have a much more deleterious effect than some of those drawbacks already mentioned. The trouble is that a lot of spurious energy is created during transition times and significant distortion is created. It follows that when you change more bits, the likelihood of error goes up. Quite often, the worst distortion occurs at the middle of the range, when a DAC must follow digital input words changing between a one followed by a lot of zeros and a zero followed by a lot of ones. As mentioned previously, DACs suffer from several other distortions; those are worth reviewing here in the context of DDS.

DACs for DDS: The Need for Speed

DACs add their unique distortions to the output of a DDS. Those are the amplitude non-linearities and glitch production discussed in Chapter 2. Non-linearities in a DAC create harmonic distortion and IMD at its analog output stages. Remember that aliases are also present at the output of a DAC. The presence of those undesired products sometimes causes unexpected spurs to appear. Harmonics of desired output signals exceeding half the sampling frequency may fold back into one's passband of interest. For example, the ninth harmonic of a 1-kHz wave sampled at 16 kHz may appear at 7 kHz. See **Fig 7.3**. It cannot appear at 9 kHz because that is greater than half the sampling frequency. That signal transforms to its first alias since it is beyond the Nyquist limit.

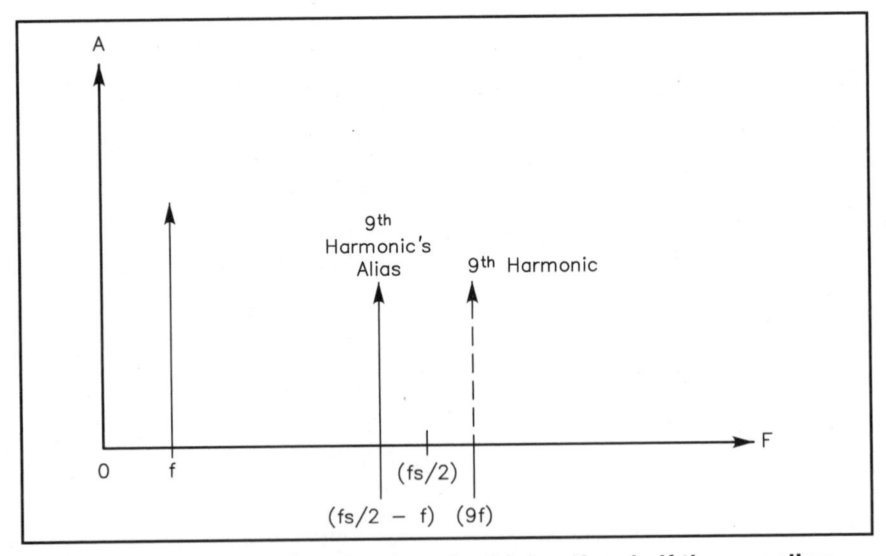

Fig 7.3—Showing that signals otherwise higher than half the sampling frequency fold back (alias) to lower frequencies.

As good as they are, the analog output stages of typical DACs are not perfectly linear. That gives rise to non-linear mixing of whatever products are present in those output stages. Usually, such mixing products are so far below one's signals of interest that they are irrelevant; however, in applications demanding maximum dynamic range, they may be significant. In addition, quantization noise produces a noise floor at the DDS and DAC output that ultimately limits dynamic range. That leads us to define what we mean by dynamic range in a DDS.

A popular way to define it is to take the ratio of the desired signal's power to that of the largest undesired product present. That may be done over the entire output bandwidth of the device (half the sampling frequency), or it may be restricted to some band of interest. In any case, this measure is called the *spurious-free dynamic range* or *SFDR* of the device. Since devices behave differently at different sampling rates, SFDR must be referred to the sampling frequency as well as the measurement bandwidth.

Fortunately, manufacturers now commonly specify the SFDR of their DACs for their intended sampling rates. Tremendous progress has been made in correcting the above-mentioned errors in DACs and some very good performance is now available. It is being discovered that as speeds increase, glitch energy becomes a more significant factor in spectral purity. Clearly, simultaneity of bit changes is harder to achieve as clock rates go up. Industry leaders have reached remarkable circuit symmetry and performance levels in this area.

DDS in Software: A Digital BFO

In many cases, DDS is used solely to generate digital sinusoidal sequences for use in subsequent digital processing. One example of that was seen in Chapter 4, where such a sequence was multiplied with the impulse response of a LPF to produce a BPF. The process in software is basically the same as in hardware: Use a phase accumulator to track phase and a look-up table to find amplitudes. One trouble with software implementations, though, may be that the size of the look-up table becomes exceedingly large. Embedded applications running on general-purpose DSP hardware, for example, may not have the luxury of megabit-sized tables. In addition, the bit-resolution of the phase accumulator and table entries may be significantly restricted.

One subject of interest is how the size of a look-up table relates to the total distortion produced by it. SFDR is a very useful measure of performance, but it actually says little about total distortion levels. It specifies maximum spurious levels relative to the desired signal, but does not specify the number of spurs or their spectral placement. An analysis of total distortion levels may begin by finding the errors between the digital representation of a wave and its analog equivalent. When the total RMS error has been computed, we may define the total distortion, in dBc, as:

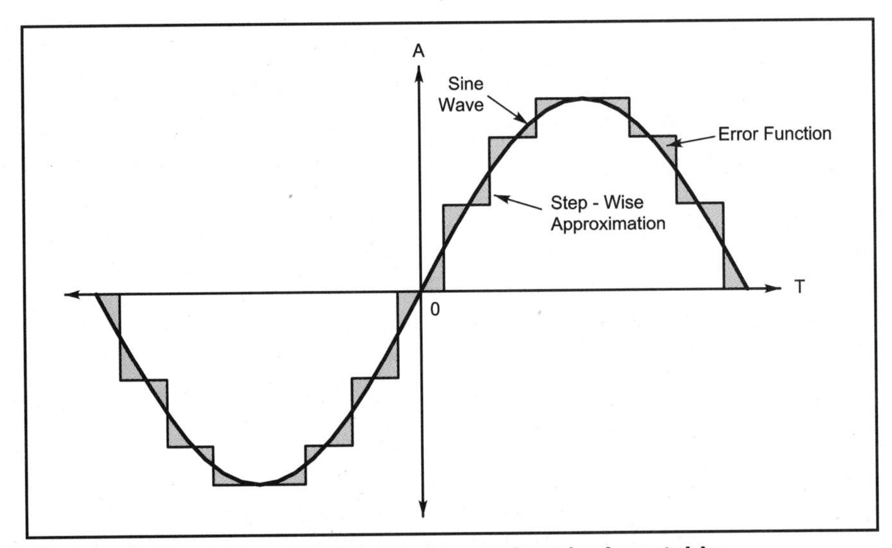

Fig 7.4—A sine wave superimposed on a short look-up table approximation of the same sine wave. The shaded areas represent errors.

Table 7.1
Total Distortion of Non-Interpolated Look-Up Tables

Length	*Total Distortion (less aliases)*
16	–20 dBc
64	–32
256	–44
1024	–56
4096	–68
16,536	–80

$$P_{TD} = 10\log\left(\frac{P_{RMS\ error}}{P_{desired}}\right) \text{dBc} \tag{97}$$

It is obvious from the nature of sampling that the largest errors occur when the function being generated has a large slope. That is, errors are large when quantization levels change a lot from one sample to the next. For a sine wave, that happens near the zero-crossing area. It follows that errors are at a minimum when the slope is near zero. That is the case for a sine wave near its positive and negative peaks. See **Fig 7.4**. The task is to find the RMS value of the shaded areas: the error function. Notice we are only interested in the long-term average of the error's RMS value. The mean-squares method of Appendix

A may be used to find that. Distortion levels calculated in this way for various table lengths are shown in **Table 7.1**.

It is perhaps interesting to note that those total distortion levels are not very close to those demonstrated for maximum spur levels under "Digital Effects" above. Both predictions take into account that all spurious energy may be concentrated at or near a single frequency, so the total may be equal to the maximum discrete level. This second prediction seems to imply that the total spurious energy is constant, even if many spurs exist at several frequencies, and over the entire range of desired output frequencies. In actual fact, Table 7.1 shows predicted distortion levels when a sine wave is being generated that exercises all the table entries the maximum number of times. Sine waves that use each table entry only once (those of higher frequency but with the same sampling rate) may exhibit lesser errors. It also shows that doubling the table length results in a 6-dB improvement.

It is very important to bear in mind, though, that when such digitally generated sequences are used in filter translation or complex mixing, distortion often falls harmlessly in frequency bands outside the bands of interest, or may tend to cancel inside the bands of interest. For example, when a complex DDS BFO is employed in a complex digital mixer, it can be shown that significant distortion is much less than what it shown in Table 7.1. That is because almost all the distortion is harmonically related to the sampling rate and therefore falls either at half the sampling frequency or at zero frequency.

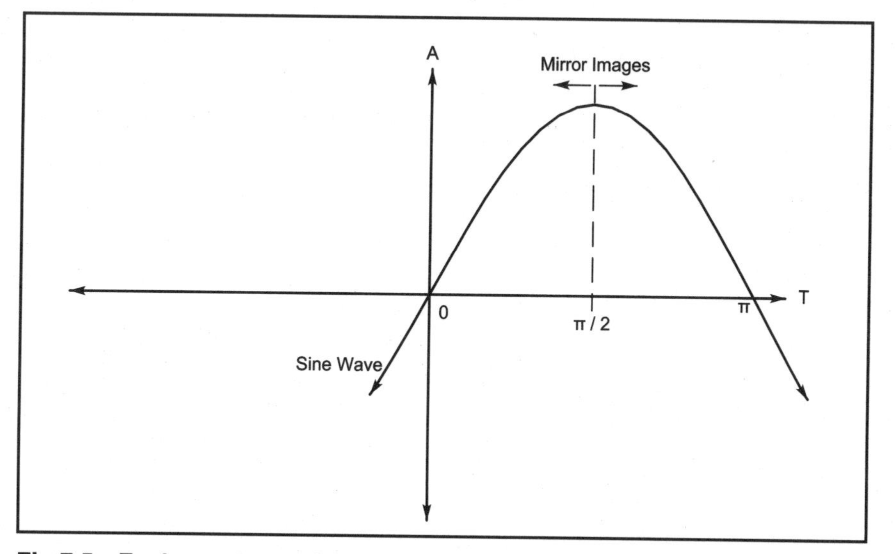

Fig 7.5—Each quarter cycle of a sine wave is the same basic shape as every other quarter cycle.

One way of reducing look-up table size involves the symmetry of sinusoids. A sine wave has a shape that repeats itself every cycle; but its basic shape is contained in each quarter cycle. The first quarter cycle looks like a mirror image of the second, and so forth. See **Fig 7.5**. With a little additional overhead, this fact may be used to reduce look-up table size by a factor of four.

In DDS, the phase accumulator may be normalized to angles within the range 0-$\pi/2$ by adjusting it by integer multiples of $\pi/2$ (with respect to the mirror imaging), then correcting for amplitude polarity at the look-up table's output. No additional distortion is created with this technique, but it does require a bit more processing time. Another way to reduce look-up table size is to employ the process of interpolation.

Look-Up Table Interpolation

Interpolation has already been defined as the adding of samples between existing samples. Interpolation of table data does exactly that; the technique has been covered in the section on square roots in Chapter 5. Now we are interested in how much reduction in distortion a straight-line interpolation gives us from those figures shown in Table 7.1. Again, the long-term average of the error function's RMS amplitude must be found. See **Fig 7.6**. That is not as easy as the non-interpolated case because the error segments have a slightly more difficult shape. Still, the problem is not intractable and **Table 7.2** shows some results. Note that when straight-line approximations are being used, interpolation factors of two and higher behave identically. In other words, all interpolated points lie on the same line. Once again, rather than find the exact solution, a triangular shape for the error segments is assumed. A calculation of the RMS value of these little triangles or sawteeth is shown in Appendix B.

Just as in Chapter 3, we could insert samples of zero between existing table entries and subsequently filter the resulting sequence to smooth it. That may be an attractive possibility for higher-order interpolation (multiple zero insertions and filtering), especially where dedicated filtering hardware is present. Only the segment of interest need be filtered, but this technique involves a heavy trade-off of table size for computation time. Besides, in the case where table size is at a premium, or a table is not possible at all, there are other ways of generating repetitive waveforms in DSP. These also fall in the DDS category by definition: They are direct synthesizers.

McLaurin Series

McLaurin series are a special case of Taylor series: power series that, in summation, converge to a meaningful result. For sine waves, we can use:

$$\sin(\phi) = \sum_{n=0}^{\infty} \frac{\phi^{2n+1}(-1)^n}{(2n+1)!} = \phi - \frac{\phi^3}{3!} + \frac{\phi^5}{5!} - \frac{\phi^7}{7!} + \cdots \tag{98}$$

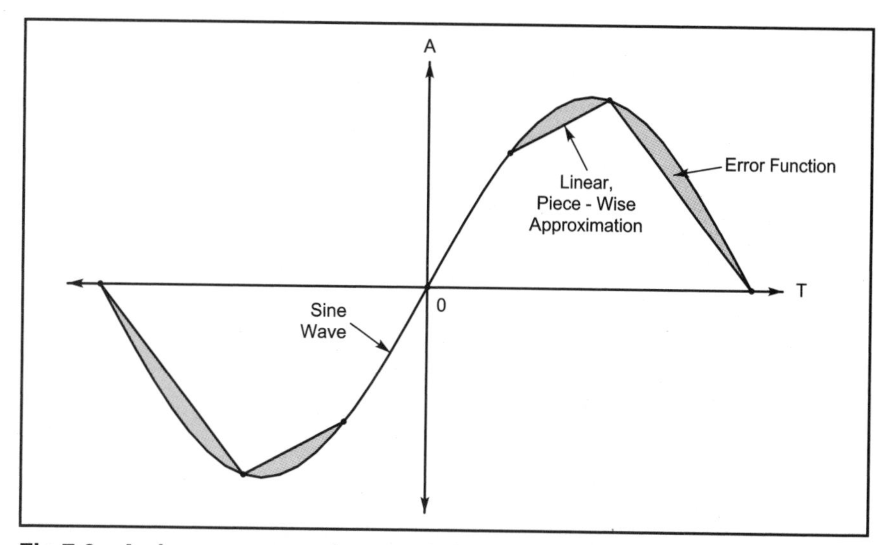

Fig 7.6—A sine wave superimposed on a first-order, linear interpolation of the same sine wave. The shaded areas represent errors.

Table 7.2
Total Distortion of Look-Up Table With 1st-Order Interpolation

Length	*Total Distortion (less aliases)*
16	–33 dBc
64	–45
256	–57
1024	–68
4096	–80
16,536	–92

For finite summations, it is again of concern how much distortion this involves since these sinusoids are so critical to DSP transceivers. To minimize the angle, ϕ, let us presume we are using the normalization procedure described above for table reduction size and that ϕ lies between 0 and $\pi/2$. The result must, of course, again be de-normalized after the computation by adjusting for polarity, if necessary.

Now we cannot compute the series forever; we can only compute several terms—and the fewer, the better. Leaving out higher-order terms and taking only the first four, for example, means that the error is equal to the sum of the terms omitted. Taking the ratio of the truncated summation, which is very close to unity at maximum, to the error at the highest angle, $\pi/2$, yields:

$$\text{error} \approx 20\log\left(\sum_{n=4}^{\infty}\frac{\phi^{2n+1}(-1)^n}{(2n+1)!}\right) = -82\,\text{dBc} \qquad (99)$$

This is very good performance and it probably exceeds the need in most applications. The error is very much less at smaller angles. A similar series is available for cosines:

$$\cos(\phi) = \sum_{n=0}^{\infty}\frac{\phi^{2n}(-1)^n}{(2n)!} = 1-\frac{\phi^2}{2!}+\frac{\phi^4}{4!}-\frac{\phi^6}{6!}+\frac{\phi^8}{8!}\cdots \qquad (100)$$

and this requires the computation of five terms to get similar performance; but the first term in this sequence is unity, which may not carry much computational overhead with it. Additionally, even integral powers of two may be convenient in processors that place the squared result in the input registers for the next squaring operation—to get the fourth power. The square may be multiplied by the fourth power to get the sixth power, and so forth. That also works to some extent with odd powers (sine sequence), but it is not quite as slick. Also, all those multiplications exact a toll on numerical accuracy, and that ultimately limits the performance of this algorithm in most instances.

Square Waves and Other Shapes

It is well known that a square wave is a very effective BFO or local oscillator. It would seem that a square wave is so much simpler to generate in DDS than a sinusoid that sinusoids should not be bothered with. That is true to some extent; but when an integral harmonic relationship does not exist between the sampling frequency and the output frequency, a problem arises.

If the sampling frequency is not an even integral multiple of an output square wave's frequency, the positive portion of the square wave cannot always have the same number of samples as the negative half. That means the output does not always have a 50% duty cycle. The split of samples per period of a square wave is liable to change from cycle to cycle in a syncopated pattern. The result is known as *phase jitter*. Maximum time jitter is plus or minus one sample, so the effect is worse, relatively speaking, at higher frequencies than at lower.

Phase deviation β may be readily computed from this effect and the amplitudes of the resulting sidebands found according to the methods in Chapter 5. Those sidebands may fall quite close to the desired signal, though, so care is required with this approach. Usually when square waves are employed in DDS, an even integral sub-multiple of the sampling frequency is chosen by design and then jitter problems are avoided. BFO values are exact and always contain an even number of samples per cycle.

A square wave acting as a local oscillator or BFO produces a slightly higher conversion gain than does a sine-wave BFO. That is because for a given peak amplitude, the square wave has the stronger fundamental content. It seems contrary to reason, but it can be shown that a square wave consists of an infinite series of odd harmonics, the fundamental component of which actually has greater peak amplitude than the square wave. See Appendix A for a proof of that.

As detailed before, a BFO or LO of one quarter the sampling frequency is very convenient because of the quadrature relation between I and Q channels of analytic signals. Sine and cosine functions must both be generated that bear a 90-degree relationship to one another. In both this and the square-wave case, harmonic distortion of the BFO is severe, but the harmonics fall harmlessly at either half the sampling frequency or at zero frequency. Further, those products cannot affect the output because competing terms cancel in the multiplications. Distortion levels indicated in Table 7.1 therefore apply, but all the distortion is outside the bandwidth of interest.

A square wave may be generated using DDS at any frequency by employing a DAC to produce an analog signal, then by squaring that signal in a voltage comparator or dedicated digital gate, then low-pass filtering. This has the benefit of eliminating all AM spurs. Alternatively, its output may be squared directly at the reference input of a PLL chip.

DSP in Indirect Frequency Synthesis: Hybrid Systems

As early as the 1980s, designers began to employ DDS as a reference for other frequency synthesis technologies, such as PLL.[52] During their young years, DDSs employing DACs suffered from much more of the distortion effects outlined above than is common today. It was thus necessary to employ methods to clean them up. A PLL has a lot to commend it for this purpose. There are also a few drawbacks.

The fine-tuning ability and superior phase-noise performance of DDS lend themselves to its use as a variable frequency reference. Recognize that a PLL may be tuned by changing its reference just as well as by changing its programming. In fact, a PLL eliminates all the spurs outside its loop bandwidth when used in that way. The main drawback is that a PLL also *amplifies* DDS spurs *inside* its loop bandwidth. The gain is simply the ratio of the VCO frequency to that of the reference. That ratio may be as high as 100 or more, so the gain may be 40 dB or more. That much degradation in spur levels may be tolerated in many cases, though, through careful design. This hybrid DDS/PLL technique is found in most modern amateur transceivers and continues to be very popular; however, its major drawback has led to increasing interest in another synthesis method: fractional-N.

DSP in Fractional-N Synthesis

Fractional-N PLLs are also indirect synthesizers that use feedback to select the output frequency. They do not necessarily employ DDS, but recent embodiments exploit certain DSP constructs that we learned about before. Fractional-N is discussed briefly in this section to illustrate how those DSP constructs may achieve another breakthrough in performance.

The impetus for fractional-N techniques is provided by the need to operate PLLs at ultra-high VCO frequencies and small step sizes.[53] Tuning a PLL in small steps means that the loop reference—the frequency appearing at the phase detector—must be low and that means the total VCO division ratio, N, is large. For example, a cellular-phone synthesizer operating at a VCO frequency of 900 MHz and which must be tuned in 10-kHz steps has N=90,000. The problem is that divider and phase-detector noise, like the reference spurs caused by a DDS in the DDS/PLL hybrid, are amplified inside the loop bandwidth by a large amount. When N=100, the amplification is 40 dB and a phase-detector noise level of –165 dBc/Hz (the noise power in a 1-Hz bandwidth) is increased to –125 dBc/Hz. That is still an excellent figure and the problem is masked by other noise sources. When N=90,000, though, the gain factor is 20 log (90,000) or almost 100 dB; that –165 dBc/Hz would be increased to –65 dBc/Hz, which is quite poor.

One relatively simple way to reduce N is to increase the loop reference frequency, but that also increases the tuning step size by an identical factor. In the 1980s, *dual-modulus* or fractional-N dividers became available on PLL synthesizer chips. That made it possible to make the division ratio, N, something other than an integer, thus achieving finer tuning steps. During a series of reference periods and phase comparisons, a dual-modulus divider divides by N for several iterations and then divides by N+1 for the next several iterations. The average division ratio is then part way between N and N+1 and may be said to be fractional. During a sequence of L reference periods, if we divide K times by N+1 and L-K times by N, the average division ratio is:

$$\frac{K(N+1)+N(L-K)}{L} = N+\frac{K}{L} \tag{101}$$

which places the fractional part, K/L, under our control. L is known as the *fractionality* of the circuit. The reference frequency may now be L times higher than before while retaining the same step size. The transfer function so created is a step-wise curve, though; the phase detector output contains phase jitter at the rate the division ratio is changing. The jitter, in turn, makes PM sidebands on the VCO unless we can filter them. Since that is not always very practical, a couple of ways to compensate the problem have been devised: analog (second-order) and digital (third-order) compensation.

A second-order, analog correction scheme applies an artificial phase advance to the phase detector when the division ratio changes, thereby compensating the step function in the transfer curve. Such compensation is imperfect: Dual-modulus spur levels of 35-40 dB down are common and rather complex external loop filtering is still required to meet the 70-dB-down levels usually sought. Experimenters have now worked out a digital correction scheme that relies heavily on DSP sampling theory. *All-digital fractional-N* largely solves the dual-modulus problem.

A third-order sigma-delta modulator is employed that effectively samples the division and ratio-changing process at a rate very much higher than the loop's reference.[54] As discussed in Chapter 2, that spreads the divider and phase-detector noise over a much larger bandwidth; most of the noise then falls outside the loop bandwidth of the PLL, and so is eliminated. Changes in the division ratio are smoothed out and dual-modulus spurs are also therefore reduced. The fractionality, L, is greatly increased and the frequency resolution made very fine. To do this, though, the phase detector must operate at a much higher speed, and that points to a drawback.

While the gain in phase-noise performance is 20logM, where M is the oversampling ratio of the sigma-delta modulator, phase detectors tend to lose noise performance at a rate of about 3 dB per octave of reference frequency. A net gain in performance of about 10logL is the result. This is definitely worthwhile: When M=16, the gain is 10log16 or about 12 dB. At the time of this writing, though, commercially available, all-digital fractional-N synthesizers have not quite attained that level. Further, the chips are not yet seen on the market by themselves and manufacturers have held the technology fairly close to their vests. Perhaps the 21st century will see these devices invading new designs. While adding a great deal of internal complexity, such a synthesizer chip eliminates a lot of external stuff and an entire analog synthesizer may be made physically quite tiny.

Interference Reduction

Although adaptive filtering and Fourier transforms are discrete fields of study in DSP, they are presented together here because they each uniquely display elements of the same, very important concept: the equivalence of time-domain and frequency-domain representations of signals. Separate yet equal ways of representing a signal are possible. The choice of a representation is often made with regard to what advantages it brings to a particular system. In many cases, selection is also influenced by what data acquisition means are available and certainly by the function to be performed. In this chapter, we will examine how spectral analysis is derived from time-domain properties of signals and how it is put to use.

Adaptive Filtering

The term adaptive means that system parameters are allowed to change under control of some algorithm to achieve a goal. That idea has far-reaching implications, not only in the DSP world, but across the spectrum of human endeavor. It is analogous to a learning process. Each of us who has learned to drive, for instance, has discovered that an automobile does not always go perfectly straight down the road. Many little corrections are necessary to keep the thing in one lane– well, at least some of us have learned that! But for certain other useful analogies, though, the discussion below is restricted to those techniques that help transmit and receive communications.

Among other things, we shall look at how the coefficients of a DSP filter may be altered on the fly to segregate repetitive signals from non-repetitive signals. A CW signal is one example of a repetitive signal; noise is the most common example of a non-repetitive signal. In this case, the repetitive signal is what we want to extract and the non-repetitive signal is to be suppressed. In other cases, we may want to suppress the repetitive signal (an interfering tone) and keep the non-repetitive (a voice signal). It is just a matter of how we characterize things and how we set up system parameters.

Many of the constructs described here have found their way into production equipment, but they are still not generally well understood. The goals of noise and interference reduction provide a logical starting point for their consideration.

Adaptive Noise-Reduction Techniques

The nature of information-bearing signals is that they are in some way coherent; that is, they have some feature that distinguishes them from noise. For example, voice signals have attributes relating to the pitch and pace of a person's speech. CW signals are perhaps the simplest example since they constitute only the presence or absence of a single frequency.

A lot of research has been done about detection of a sinusoidal signal buried in noise.[55] Adaptive-filtering methods are based on the exploitation of the statistical properties of a desired signal. Specifically, they analyze how past samples of the signal relate to current samples in the time domain. The measure of their similarity is called *autocorrelation*.

Autocorrelation must be determined with respect to some time offset. Example: If you examine a 1-Hz sine wave and measure its autocorrelation at a 1-second offset, you will get a value of one or 100%: That sine wave repeats exactly every second. If you use a time offset other than an integral multiple of one second, you will get an autocorrelation value of less than unity, since samples of the sine wave do not exactly match at each sample time. Now it follows that if you are trying to find a 1-Hz sine wave buried in noise, you should compare samples that are one second apart and see what kind of match you get. It is evident that the autocorrelation of random noise is undefined, because no information exists about what a noise signal will do at the next sample time. This fact may be used to build what is known as an *adaptive interference canceler*.[56]

A Manually Tuned Adaptive Interference Canceler

Imagine we have digitally sampled some noisy input signal x and we want to filter it to enhance its sinusoidal content. x_t is just a sample of continuous signal x taken at time t. Let us say that the desired part of the signal is some CW. In this case, all that is required is a band-pass filter centered on the expected frequency of the CW. We know the output will take the form of a sinusoid and that only its amplitude will change.

So we set up an FIR filter structure and set the initial filter coefficients h_k to zero. Then we set up an error-measurement system to compare a sine wave, d_t, with the output of the filter, which we call y_t. See **Fig 8.1**. The *reference*, d_t, is of the same frequency we expect the CW note to be. Difference output, e_t, is known as the *error signal*. Further imagine some person who is watching the error signal on an oscilloscope and who has his or her hands on the filter controls.

Fig 8.1—A manually tuned adaptive modeling system.

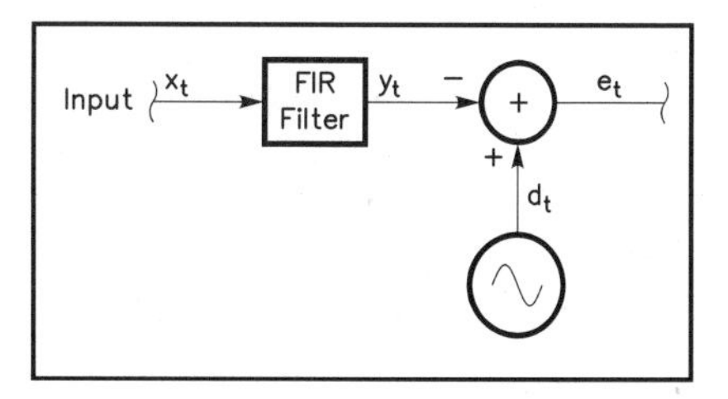

That person knows that if he can minimize the error, the filter will have converged to a band-pass filter centered on the frequency of d_t.

The speed and accuracy of convergence is dependent on how well the person can analyze the error data. If it is hard to tell a CW signal is present, then adjusting the filter correctly will also be difficult. Further, the person can only make measurements at a rate determined by his individual talents. He can check the error only so often, or he can arrange to make long-term averages of it. Like a driver with a learning permit, the person discovers that if he turns the controls the wrong way, the error increases. That feedback is used to reverse the direction of adjustment and the person will turn the controls the other way. The person soon discovers that he is on a *performance surface* that has an "uphill" and "downhill". The pre-defined goal is to go only downhill.

So thrashing about with the controls continues. Sometimes mistakes are made, but fortunately, no call to an insurance agent is necessary! Ultimately, the person makes headway down the hill and at some point, the error gets very small. The person knows he is near the "bottom of the bowl." Once at the bottom, it is uphill no matter which way he turns. The goal of minimizing the error has been achieved. This story is analogous to aligning a regular band-pass filter with an adjustment tool.

After doing the whole thing several times, the person finds that certain rules help speed up the process. First, he finds there is a relationship between the magnitude of the error signal and the amount of adjustment that must be made to the controls. When the total error is large, a large amount of tweaking must be done; when small, it is better to make only small adjustments to stay near the bottom of the performance surface. Second, he finds there is some correlation among the error signal, the input signal and the coefficient set he is adjusting.

Derivation of algorithms that provide for steepest descent down that hill is a long and tedious exercise in linear algebra. The person goes to the library and discovers that two fellows named Widrow and Hoff figured it all out around 1960.[57] They state you should make adjustments at each sample time t according to:

$$h_{t+1} = h_t + 2\mu e_t x_t \qquad (102)$$

where h is the entire filter coefficient vector. This is the *least mean squares (LMS)* algorithm.

Properties of an Adaptive Interference Canceler

This algorithm transforms the adaptive modeling system of Fig 8.1 into the adaptive interference canceler of **Fig 8.2**. Notice that both the desired and undesired outputs are available. This is nice in case we want to accept the incoherent signals instead of the coherent ones. Such is the situation for an adaptive notch filter, treated further below; however, the algorithm does not have to change to do that. Performance issues of interest with this circuit are the amount of adjustment error in the steady state and the speed of adaptation.

One of the first things evident about the LMS algorithm is that the speed of adaptation and the total misadjustment are both directly proportional to μ. Its value, which generally ranges between zero and one, is chosen to set the desired properties. Note that we have a trade-off between speed and misadjustment. Large values of μ result in fast convergence, but large steady-state adjustment errors. The length of the filter also has a bearing on both those parameters. Total misadjustment may be shown to be directly proportional to the filter's length and this may be a limitation in some systems. Also, the delay through the filter grows in direct proportion to its length. That may also be a restriction under some circumstances.

Attempts may be made to vary the factor μ on the fly to get around some of those issues. When the error is large, a large value of μ may be selected to obtain fast convergence; as the error shrinks, its value may be decreased to maintain a small misadjustment. That works fine as long as the input signal's characteristics are not changing rapidly.

Values of μ greater than unity have even been tried to get very fast convergence, but the LMS algorithm is not necessarily stable then. The inventors refer to that arrangement as the "dangerous LMS algorithm." Enough said.

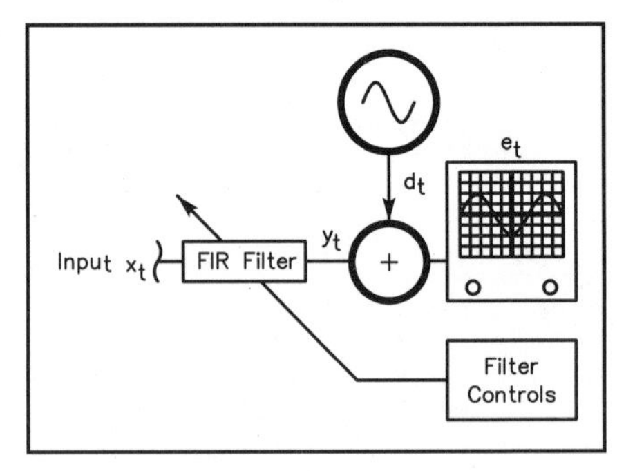

Fig 8.2—A manually tuned adaptive interference canceler.

An Adaptive Interference Canceler Without an External Reference: An Adaptive Predictor

In the example above, we knew pretty much what we were looking for: a CW signal of known frequency. What happens when we do not have such clues about a signal's information content ahead of time, except that it is somehow repetitive? Many situations like that arise in practice. It might seem at first that adaptive filtering could not be applied. But if a delay, z^{-n}, is inserted in the *primary input*'s path (that of x_t) to create the reference signal, d_t, periodic signals may be detected and therefore enhanced or eliminated. See **Fig 8.3**. This delay is just the autocorrelation offset mentioned before. It represents the time offset used to compare past input samples with current ones. The delay must be chosen so that desired components of x_t correlate and the undesired components do not.

This system is an *adaptive predictor*. **Fig 8.4** shows the result of an actual experiment using a sine wave buried in noise as the input. Input bandwidth is 3 kHz and the input SNR = 0 dB. For any given value of μ, the filter converged quickest on the solution when the delay was set to roughly the amount of delay through the filter. Note that the filter's impulse response is also a sinusoid. The filter's bandwidth was measured and was always close to:

$$BW = \frac{2\mu A^2}{t_s} \tag{103}$$

where A is a long-term average of the peak amplitude of the input, x_t, and t_s is the sampling period. So we find that the speed of adaptation and bandwidth–hence, noise-reduction effectiveness–are both proportional not only to μ, but also to the amplitude of the input signal. In Fig 8.4's example, μ=0.005, A=1 and t_s 60 μs. The SNR improvement is the ratio of pre-filtered bandwidth to final bandwidth, or:

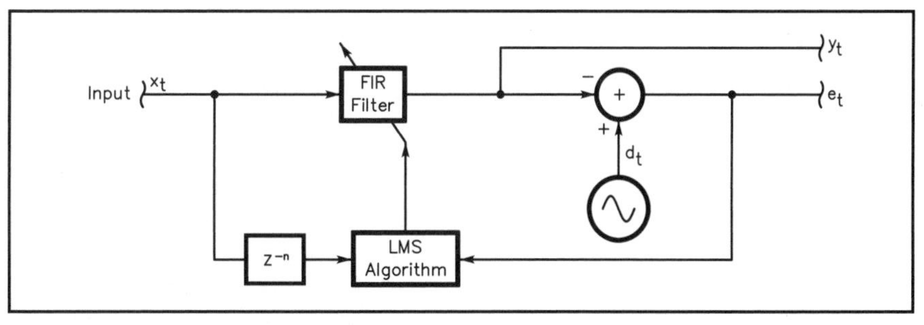

Fig 8.3—An automatic interference canceler or adaptive predictor.

Fig 8.4a—A sine wave buried in noise (SNR = 0 dB).

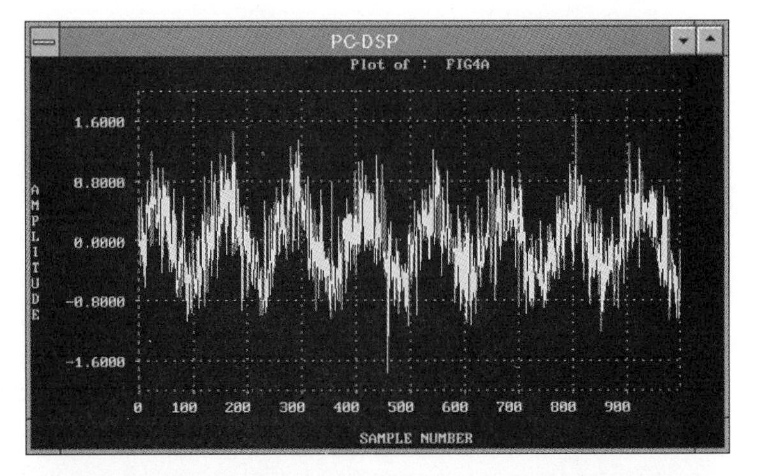

Fig 8.4b—Sine wave after noise reduction.

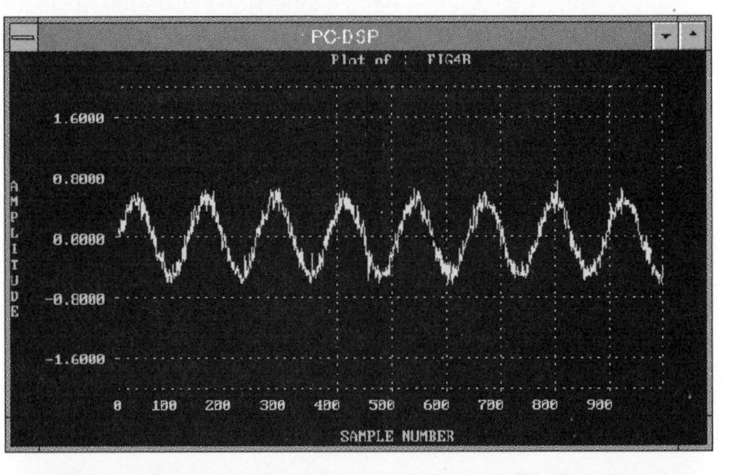

Fig 8.4c—The adaptive filter's impulse response. The impulse response of an FIR filter is the sequence of multipliers used in its taps.

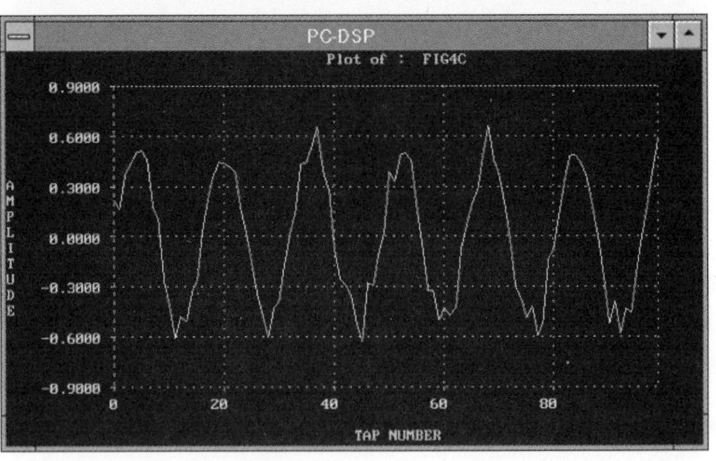

$$\Delta SNR = 10\log\left(\frac{3000}{BW}\right)$$

$$= 10\log\left(\frac{3000t_s}{2\mu A^2}\right) \tag{104}$$

$$= 13 \text{ dB}$$

To create an adaptive notch filter, the unpredictable components e_t may be taken as the output. Say we have a voice signal being corrupted by the presence of a single interfering tone or carrier–a very common situation on today's HF bands. We may set z^{-n} and μ so that the tone is predictable and the voice signal is not. The filter will converge to the solution that removes the tone and leaves the voice signal virtually unscathed. The bandwidth of the notch is the same as Eq 103, but its theoretical depth is dependent only on numerical accuracy effects in the processing system. In actual practice, notch depth is usually determined by misadjustment error, as described.

When adaptive filters with many taps are used, multiple tones may be notched. In fact, Widrow and Stearns (see Reference 56) have shown that as few as two taps for each tone may be used. In this case, though, filter coefficients do not necessarily approach any fixed value but may "roll" with the input signal.

Refer to **Fig 8.5**. This circuit is basically the same as that of Fig 8.2, except that only two taps are used in the filter and a complex reference is employed: one that has both a sine and cosine part. The thing is capable of manually notching a single tone at the frequency of the complex reference. It may be shown that this system is linear and time-invariant for the output, e_t.

In this chapter's analogy of a person trying to go down the hill, it was shown that if it is hard to tell a repetitive signal is present, it may be difficult to adjust filter coefficients to enhance or eliminate it. Remember that the key notion driving the LMS algorithm is that of minimizing output energy. That is why adaptive notch filters are generally superior to manually tuned types: Even when the input phase of the offending tone or tones is changing, the filter tracks it to maintain a best fit. It follows that anything helping minimize output energy makes it easier to process error data. One reasonably simple concept that does it is called *leakage*.

The Leaky LMS Algorithm

A unique feature of *leaky LMS algorithms* is a continual nudging of filter coefficients toward zero. The effect of leakage may be striking, especially as applied to noise reduction of voice signals. SNR improvements increase because filter coefficients tend toward lower throughput gain in the absence of repetitive input signals. More significantly, leakage helps the filter adapt under low SNR conditions–the very conditions when noise reduction is needed most.

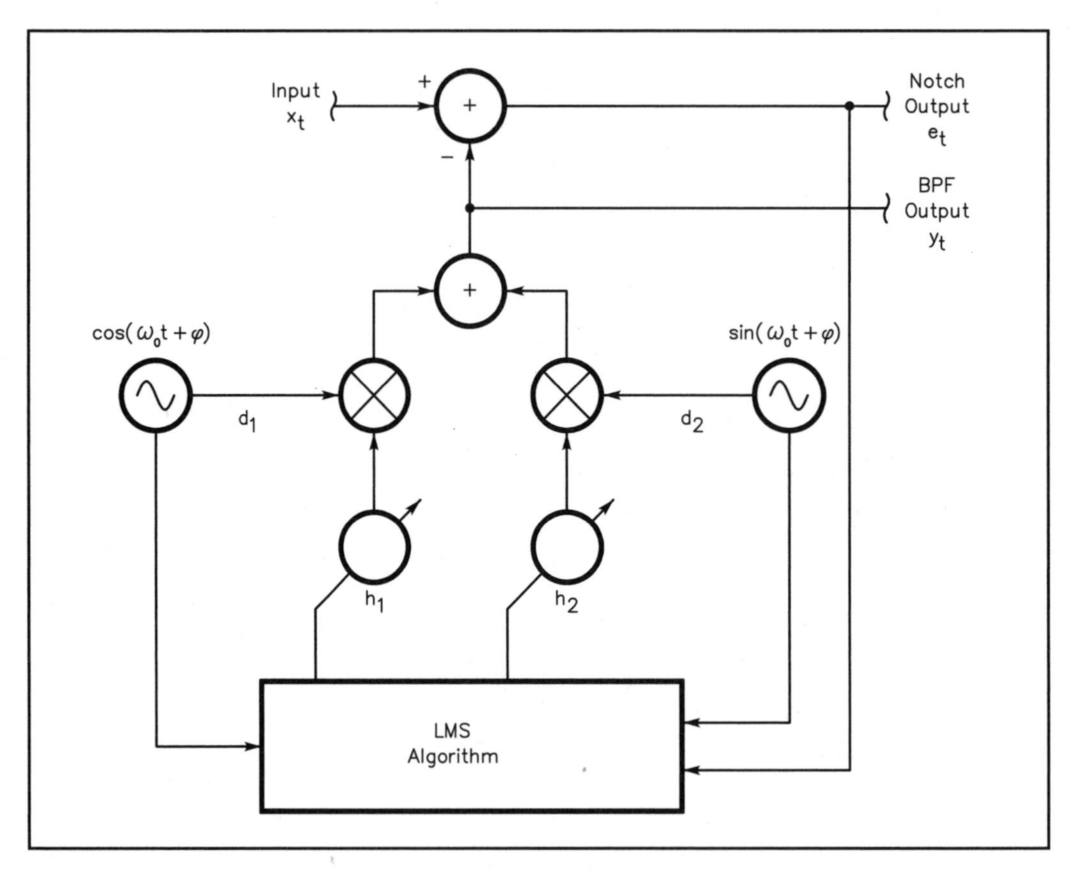

Fig 8.5—A manually tuned, adaptive notch filter. This filter can notch one frequency only.

One way to implement leakage is to add a small constant of the appropriate sign to each coefficient at each sample time. The constant is positive for negative coefficients and negative for positive coefficients:

$$h_{t+1} = h_t + 2\mu e_t x_t - \lambda[\text{sign}\ (h_t)] \qquad (105)$$

where λ is the *leakage constant.* It may be varied to change the amount of leakage. Large values prevent the filter from converging on any input components and things get very quiet indeed! Small values are useful in extending the dynamic range of the system. In the absence of coherent input signals, coefficients linearly move toward zero. During convergent conditions, the total misadjustment is increased, but this is not usually serious enough to affect received signal quality.

Alternatively, coefficients may be scaled by a factor of less than unity at each sample time, thus also nudging them toward zero:

$$h_{t+1} = \gamma h_t + 2\mu e_t x_t \tag{106}$$

where γ is the leakage constant. For values just less than one, leakage is small; values near zero produce large leakage and again, the filter is prevented from converging on anything. This realization results in a geometric decay of coefficients toward zero, which may sometimes be more desirable than a linear progression.

Note that the leaky LMS algorithm must adapt to survive, much as a hummingbird must flap its wings. Were the factor μ suddenly set to zero, all the coefficients would go to zero, never to recover. It is therefore especially dangerous to use these algorithms with adaptive values of μ.

Finally, it may be shown that the leaky LMS algorithm produces an effect that is equivalent to adding normalized noise power to the signal input equal to:

$$\sigma^2 = \frac{1-\gamma}{2\mu} \tag{107}$$

It is perhaps strange to think that anything adding noise power to the input can help reduce the noise at its output; but it is the kind of noise that increases the thrashing about of the filter controls and not the kind that worsens final SNR. It thus helps us down the hill.

Fourier Transforms

While Fourier transforms are not used exclusively for interference reduction, we present them in that light here because they are generally superior to adaptive-filtering algorithms in that application. The penalty for this greater effectiveness is an increased computational burden. The relationship Joseph Fourier (pronounced *foor*-ee-ay) formulated between the application of heat to a solid body and its propagation has direct analogy to the behavior of electrical signals as they pass though filters and other networks. The laws he wrote define the connection between time- and frequency-domain descriptions of signals. They form the basis for DSP spectral analysis, which makes them extremely valuable tools for many functions, including digging signals out of the noise, as we will see below.

A Fourier transform is a mathematical technique for determining the frequency content of a signal. Applied to a signal over some finite period of time, it produces an output that describes frequency content by assuming that the section of the signal being analyzed repeats itself indefinitely. Of course, when we analyze a real signal, such as a couple of seconds of speech, we know that those few seconds do not, in fact, repeat endlessly. So at best, the Fourier transform can give us only an approximation of the frequency content. If we looked at a large enough chunk of the signal, that approximation would be pretty good. Certain mathematical conditioning of the input data will help us control the error, even for relatively short analysis intervals.

Originally, the Fourier transform was developed for continuous signals. In DSP, we use a variant of it called the *discrete Fourier transform (DFT)*. It is the discrete version because it operates on sampled signals. It is a *block transform* because it converts a block of N input samples into a block of N output *bins*. The input block may be any N contiguous samples. A DFT makes use of complex sinusoids and produces a complex result. When the input data are real, meaning they lack an imaginary part, half the output block consists of the *complex conjugates* of the other half, and so is redundant. When a complex input is used, none of the output bins is redundant.

We saw before that a complex sinusoid is just a pair of waves: a cosine wave and sine wave together of the same frequency. Since we will be dragging around a lot of these in the equations below, recall a little mathematical shorthand for them called the *Euler identity*:

$$e^{j\omega t} = \cos \omega t + j\sin \omega t \quad (108)$$

where e is base of natural logarithms. We will shorten this even more later. For each output bin k, where $0 \le k \le N-1$, a DFT is computed as:

$$X_k = \sum_{n=0}^{N-1} x_n e^{\frac{-j2\pi nk}{N}} \quad (109)$$

Expanding Eq 109 using the Euler identity yields:

$$X_k = \sum_{n=0}^{N-1} x_n \cos\left(\frac{2\pi nk}{N}\right) - j\sum_{n=0}^{N-1} x_n \sin\left(\frac{2\pi nk}{N}\right) \quad (110)$$

So each bin has a real part and an imaginary part. Note that each part is calculated using the same kind of convolution sum we saw in previous chapters. Eq 110 is in normal complex-number form: a+jb. Coefficients a and b yield the amplitude and phase of the signal x_t at frequency f_k:

$$A_k = \left(a_k^2 + b_k^2\right)^{\frac{1}{2}} \quad (111)$$

$$\phi_k = \arctan\left(\frac{b}{a}\right) \quad (112)$$

k is directly proportional to the frequency of its bin according to:

$$f_k = \frac{kf_s}{N}, \text{ for } k < \frac{N}{2} \quad (113)$$

The bins are evenly spaced in frequency by the amount $f_1 = f_s/N$, but there are actually only N/2 real frequencies represented. As mentioned above, half the DFT bins produced from a real input are redundant. Complex inputs may analyze positive and negative frequencies separately.

Working in reverse, we may reconstruct time-domain signal x_t by summing X_k for all values of k:

$$x_t = \frac{1}{N}\sum_{k=0}^{N-1} x_k e^{\frac{j2\pi kt}{N}} \tag{114}$$

This is the *inverse discrete Fourier transform (IDFT or DFT⁻¹)*. It is important to note the duality of the DFT/IDFT relationship. The transforms are not really altering the signal in any way, they are only different ways of representing it mathematically. The strength of the DFT in noise-reduction systems is that it evaluates the amplitude and phase of each frequency component to the exclusion of others. As far as we can reduce the *resolution bandwidth*, f_s/N, we can eliminate additional noise by artificially zeroing frequency bins not meeting a pre-defined amplitude threshold. Finer resolution bandwidth is obtained by increasing the number of bins, N, decreasing the sampling frequency, or both. Increasing the number of bins, N, involves taking a larger block of N input samples; the larger block represents a longer time span. Obviously we have to wait for N samples to be taken before we can Fourier transform a complete block: A delay of N samples is the result.

Since the DFT assumes the input block repeats indefinitely, we have discontinuities at the beginning and end of the block where the data have been chopped out of the continuous string of input samples. These abrupt discontinuities cause unexpected spectral components to appear, just as fast on-off keying of a CW transmitter does. This phenomenon is known as *spectral leakage*. Discrete signal components in the input leak some of their energy into adjacent frequency bins, smearing the spectrum slightly. Increasing the number of bins, N, helps alleviate this problem. Increasing N moves the bins closer together; a signal that falls between two bins will still cause leakage into adjacent bins, but since the bins are closer together, the spread in frequency will be less. Even so, input components are still spreading their energy over several bins and this overlap makes it difficult to determine their exact amplitudes and phases. See **Fig 8.6**.

To minimize that problem, we use a technique known as *windowing* on the input data prior to transformation. The data block is multiplied by a *window function*, then used as input to the DFT normally. Window functions are chosen to shape the block of data by removing the sharp transitions in its envelope. Examples of window functions and their DFTs are shown in **Fig 8.7**. The rectangular window is equivalent to not using a window at all, as all the samples are multiplied by a constant. The other window functions achieve various

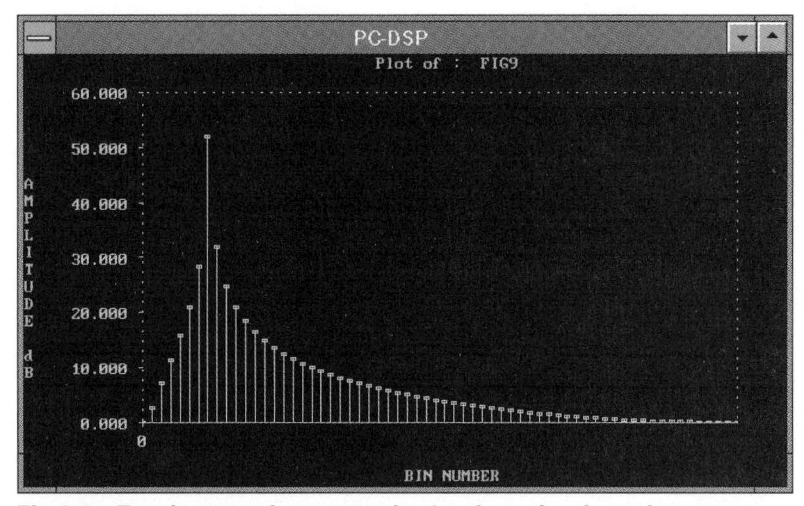

Fig 8.6—Fourier transform magnitude of a noise-free sine wave, showing spectral leakage.

amounts of side-lobe reduction. These window functions are also used to design filters using the Fourier transform method. In fact, these sequences can be used as the impulse responses of prototype LPFs, as should be evident from their frequency responses. They each involve a trade-off between transition bandwidth and ultimate attenuation. Note that in the figure, values of ultimate attenuation are plotted without regard to dynamic-range limitations which may be imposed by the bit-resolution of actual systems.

Fast Fourier Transforms

In the years before computers, reduction of computational burden was extremely desirable. Many excellent mathematicians, including Runge,[58] applied their wits to the problem of calculating DFTs more rapidly than the direct form of Eq 109. They recognized that the direct form requires N complex multiplications and additions per bin and that N bins are to be calculated, for a total computational burden proportional to N^2. The first breakthrough was achieved when they realized that the complex sinusoid $e^{-j2\pi kn/N}$ is periodic with period N, so a reduction in computations is possible through the symmetry property:

$$e^{\frac{-j2\pi k(N-n)}{N}} = e^{\frac{j2\pi kn}{N}} \tag{115}$$

This led to the construction of algorithms that effectively break any N DFT computations of length N, into N computations of length $\log_2 N$. Thus, the computational burden was reduced to be proportional to $N \log_2 N$. Because even this

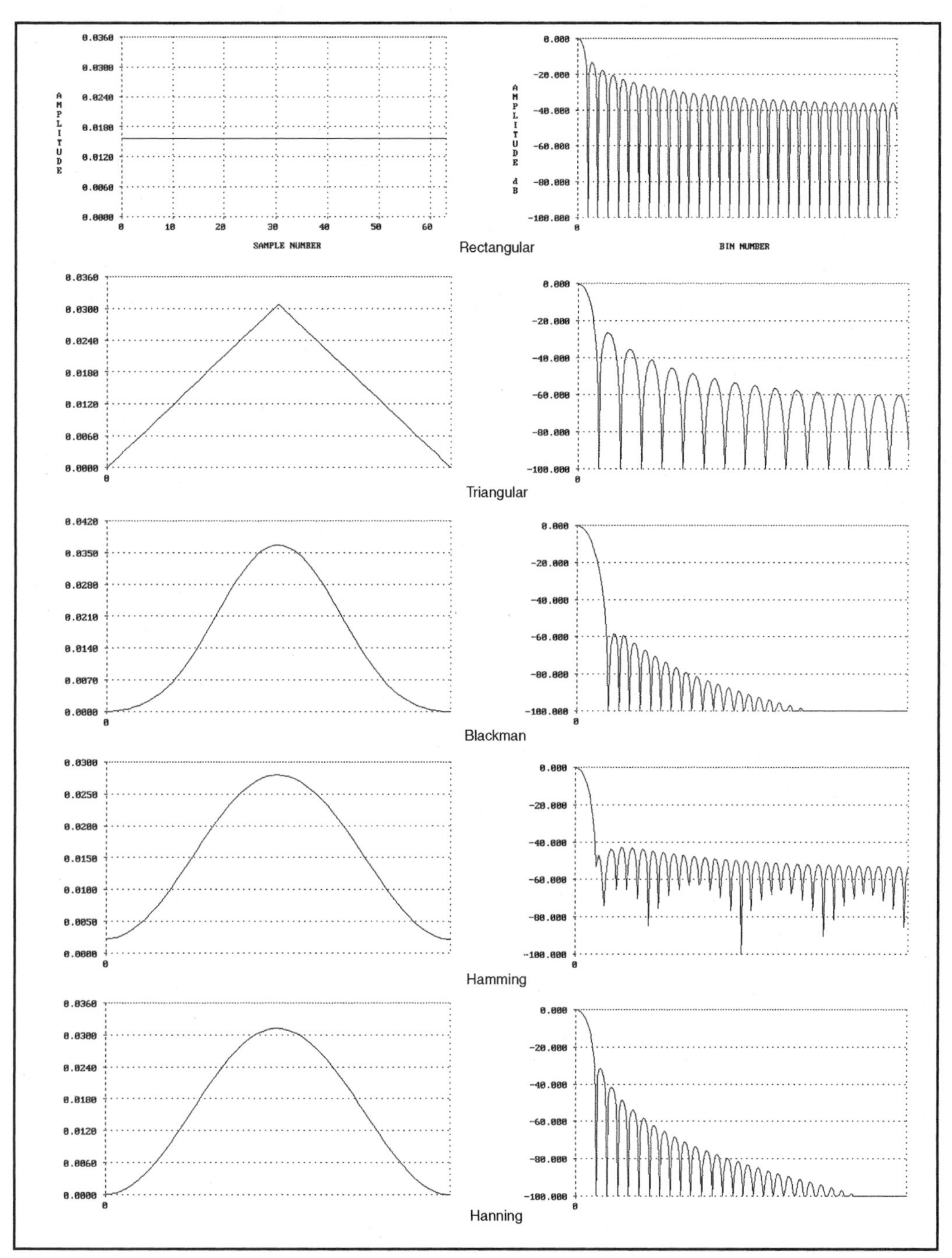

Fig 8.7—Some window functions and their frequency responses (Fourier transforms).

much calculation was not practical by hand, the usefulness of the faster algorithms was overlooked until Cooley and Tukey picked up the notion in the 1960s (see Reference 4).

To exploit the symmetry referred to, we have to break the DFT computations of length N into successively smaller calculations. This is done by *decomposing* either the input or output sequence. Algorithms wherein the input sequence, x_t, is decomposed into smaller sub-sequences are called *decimation-in-time* FFT algorithms; output decompositions result in *decimation-in-frequency* FFTs. See **Fig 8.8** for the first step in a decimation-in-time decomposition. Decomposition is based on the fact that for some convenient number of samples, N, many of the sine and cosine values are the same and products can be combined prior to computing the convolution sums. In addition, other products have factors that are other sine and cosine values. It turns out that electing to decompose by successive factors of two produces a very compact and efficient algorithm: a *radix-2* FFT algorithm.

Now for that additional bit of complex-sinusoidal shorthand mentioned earlier. Lots of complex sinusoids will appear in the diagrams to follow, so it sure would be nice to reduce the clutter a bit more. Let us follow the popular DSP text of Oppenheim and Schafer (Reference 7) and select the notation:

$$e^{\frac{-j2\pi kn}{N}} = W_N^{kn} \tag{116}$$

This is used in **Fig 8.9** in a flow chart for a complete FFT calculation, for N=8. Multiplication symbols represent complex multiplications, addition symbols represent complex additions. Note that each complex multiplication requires four real multiplications and two real additions. Complex additions need two real additions. We have eight input points and eight output points. Observe that the diagram could not be drawn without crossing many signal paths—there is a lot of calculation going on! Computations progress from left to right in $\log_2 N$ = 3 stages; each stage requires N complex multiplications and additions, so the total burden is proportional to $N\log_2 N$. Further, each stage transforms N complex numbers into another set of N complex numbers. This suggests we should use a complex array of size N to store the inputs and outputs of each stage as we go along. An examination of the branching of terms in the diagram reveals that pairs of intermediate results are linked by pairs of calculations like that shown in **Fig 8.10**. Because of the appearance of this diagram, it is known as a *butterfly computation.*

Making use of another symmetry of complex sinusoids, we can reduce the total multiplications of the butterfly by another factor of two. A modified butterfly flow diagram is shown in **Fig 8.11**. This calculation can be performed *in place* because of the one-to-one correspondence between the inputs and outputs of each butterfly. The nodes are connected horizontally on the diagram. The data from locations a and b are required to compute the new data to be

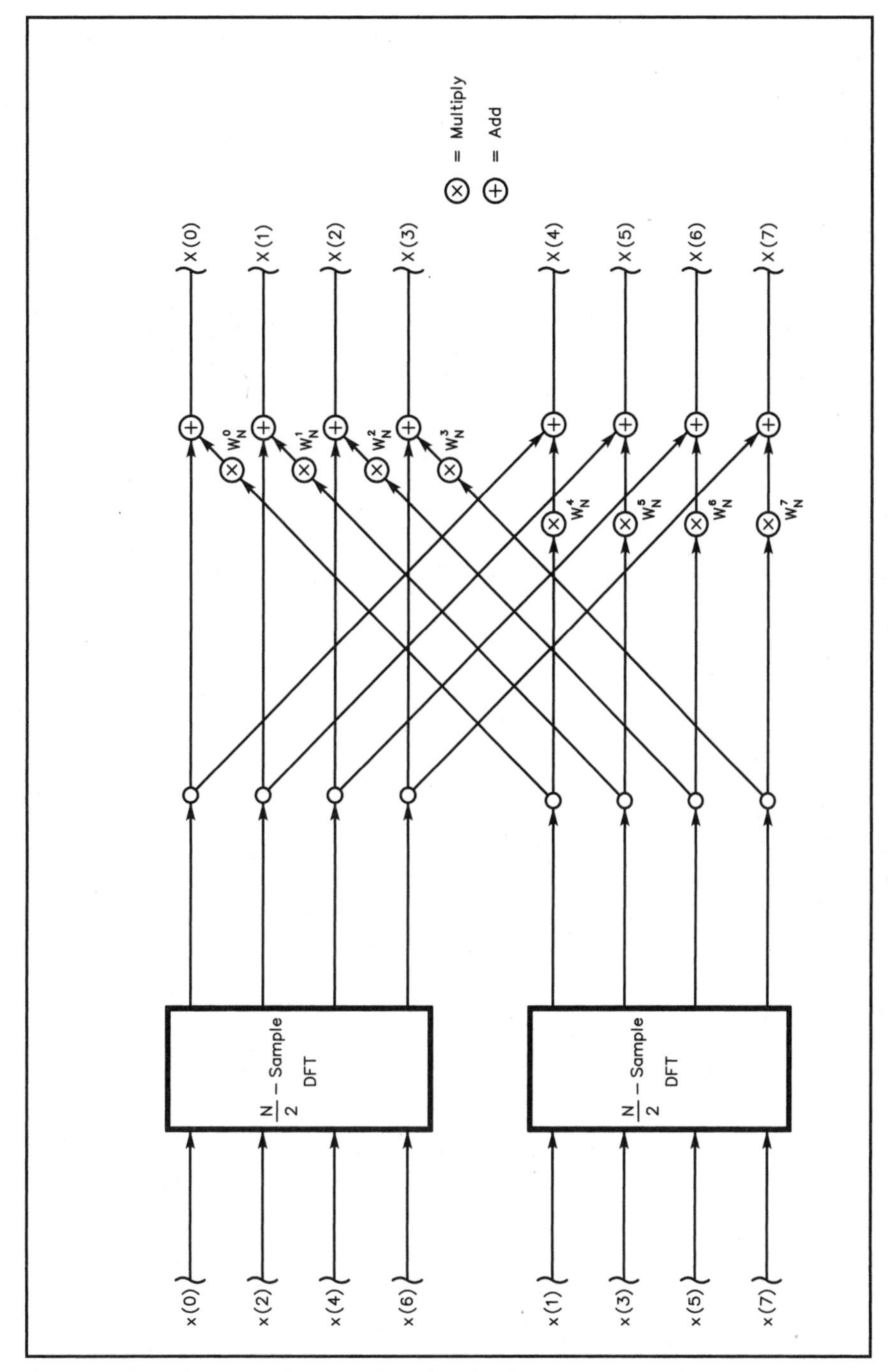

Fig 8.8—The first step in a decimation-in-time decomposition.

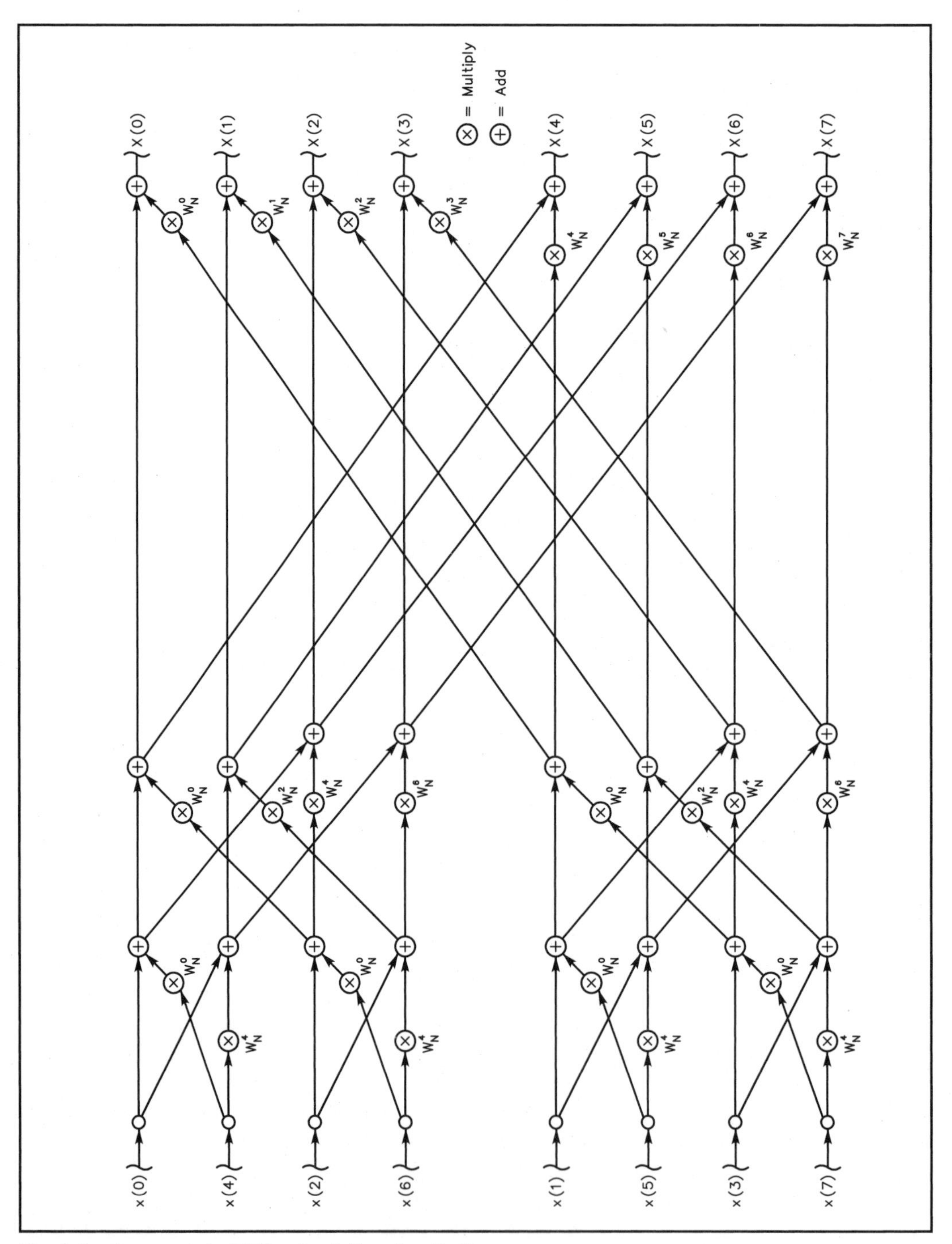

Fig 8.9—A complete FFT calculation for N=8.

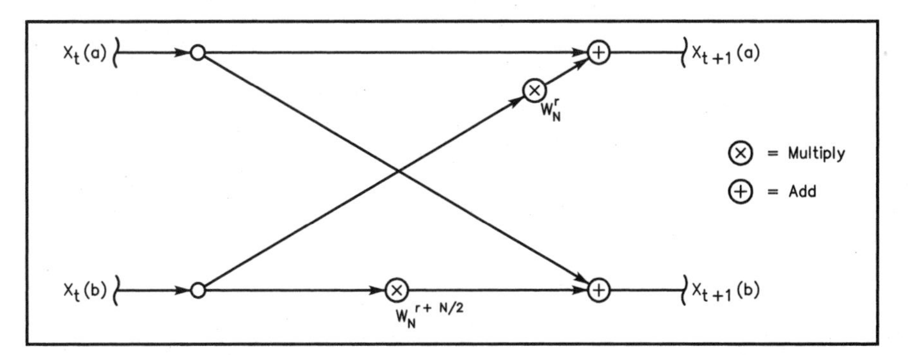

Fig 8.10—A butterfly computation.

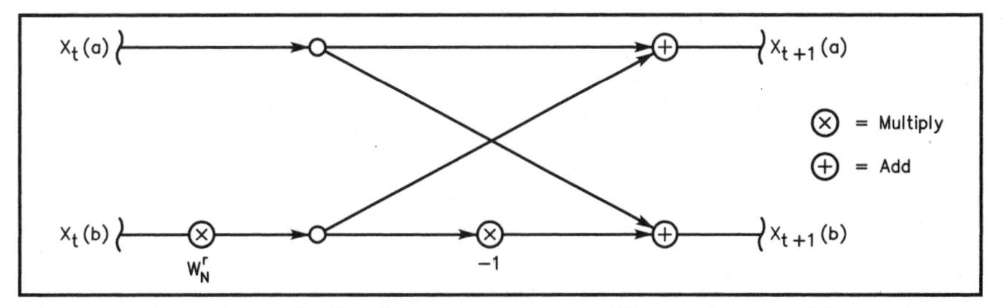

Fig 8.11—A modified butterfly computation with one less multiplication.

stored in those same locations, hence only one array is needed during calculation. A complete *eight-point* FFT with the modified butterflies is shown in **Fig 8.12**.

An interesting result of our decomposition of the input sequence is that in Fig 8.9, the input samples are no longer in ascending order; in fact, they are in *bit-reversed* order. It turns out this is a necessity for doing the calculation in place. To see why this is so, let us review briefly what happens in the decomposition process. We first separate the input samples into even- and odd-numbered samples. Naturally, all the even-numbered samples appear in the top half of the diagram, the odds in the bottom. Next, we separate each of these sets into their even- and odd-numbered parts. This process is repeated until we have N sub-sequences of length one. It results in the sorting of the input data in a bit-reversed way. This is not very convenient for us in setting up the calculation, but at least the output arrives in the correct order.

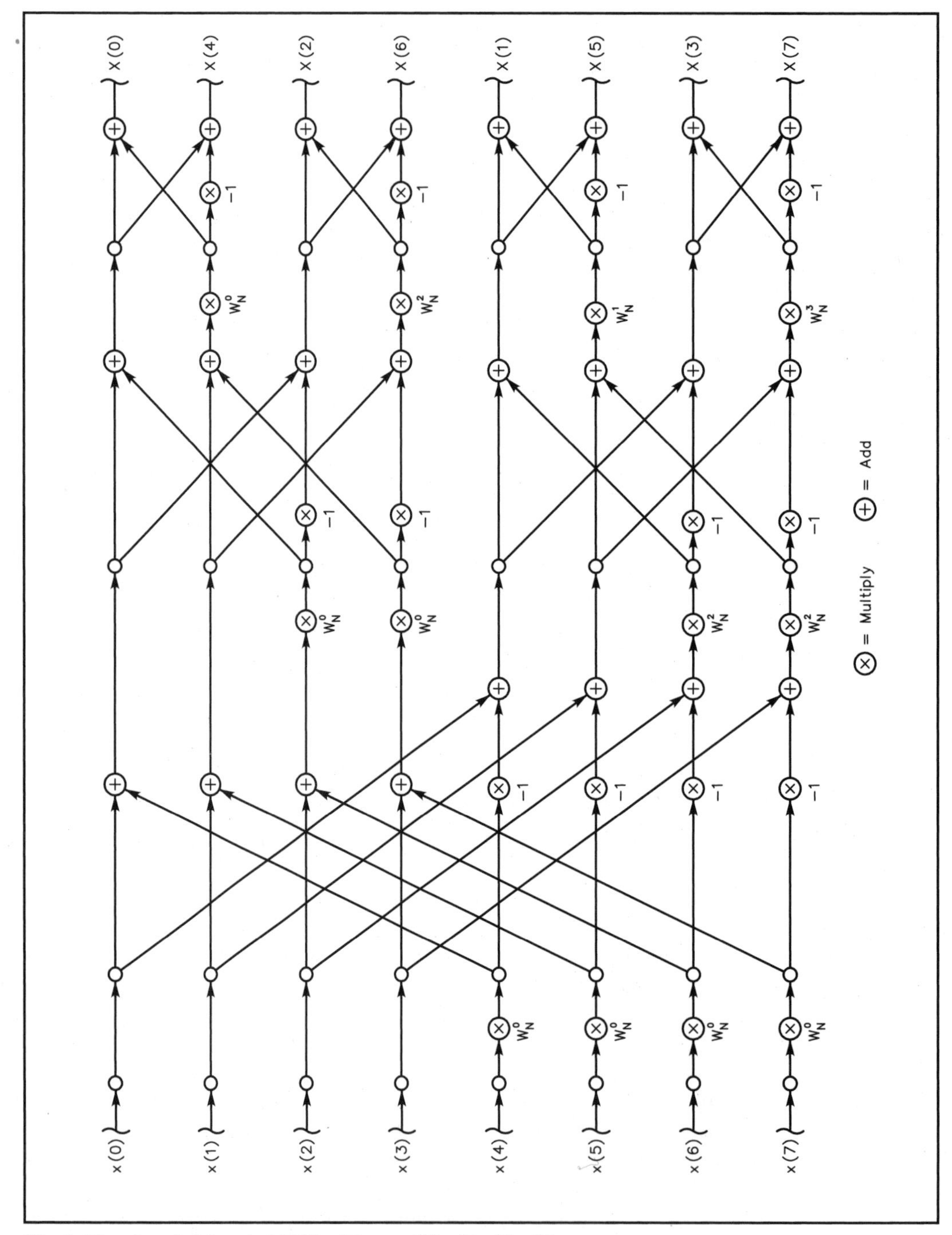

Fig 8.12—An eight-point FFT with modified butterflies.

General FFT Computational Considerations

While we are on the subject, this business of bit-reversed indexing is the first thing that ties one's brain in knots during coding of these algorithms, so let us look at ways to perform that sorting. Several approaches are feasible to translate a normally ordered index to a bit-reversed one: a look-up table, the bit-polling method, reverse bit-shifting and the reverse counter approach.

The look-up table is perhaps the most straightforward approach. The table may be calculated ahead of time and the index used as an address into the table. Most systems do not require very large values of N, so the space taken by the table is not objectionable. For more space-sensitive applications, the bit-polling method may be attractive. Since the bit-reversed indices were generated through successive divisions by two and determination of odd or even, a tree structure can be devised that leads us to the correct translation, based on bit-polling. See **Fig 8.13**. The algorithm examines the least-significant bit, then branches either upward or downward based on the state of the bit. Then the second least-significant bit is examined and another branch taken, and so forth, until all bits have been examined.

The bit-shifting method requires about the same computation time as bit-polling. Two registers are used: one for the input index shifting right through the carry bit, the other shifting left through carry. After all the bits have been shifted, the left-shifting register contains the result. See **Fig 8.14**. Finally, Gold and Rader[59] have described a flow diagram for a bit-reversal counter than can

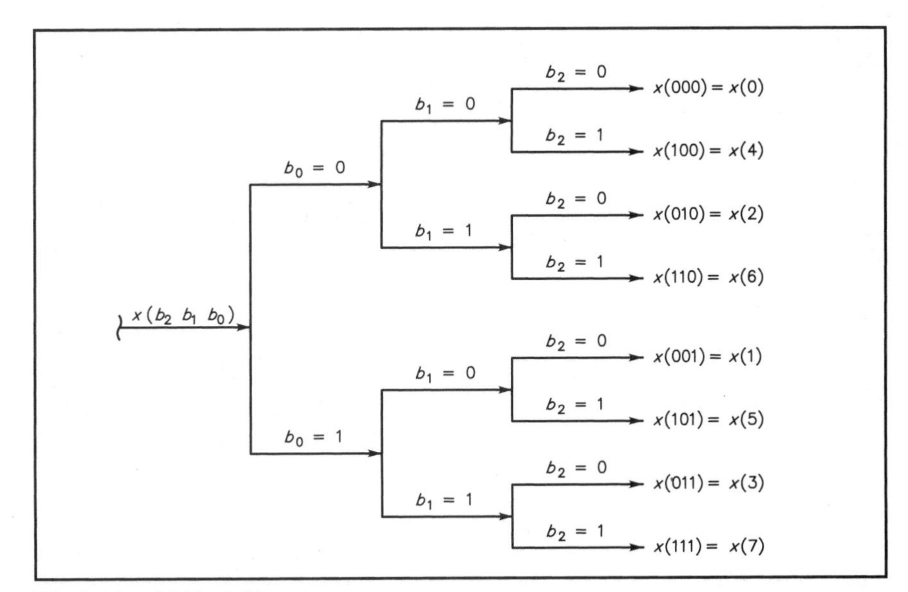

Fig 8.13—A bit-polling tree.

Fig 8.14—A bit-reversing set of shift registers.

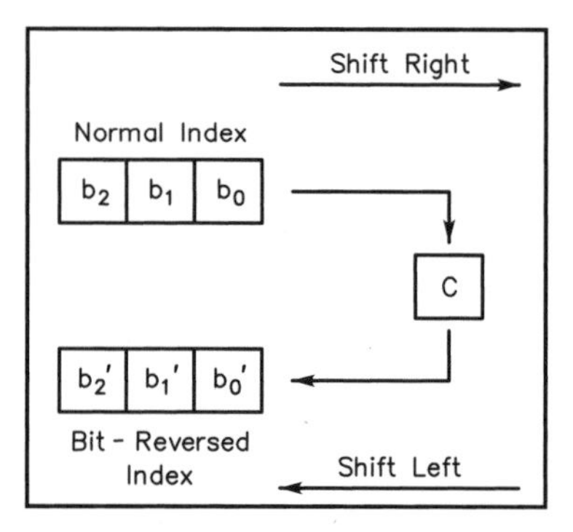

be "decremented" each time the index is to change. If data are actually to be moved during sorting, the exchange is made between data at input index n and bit-reversed index m, but only once. That is, only N/2 exchanges need be performed.

During the actual calculations, indexing of data and coefficients requires attention to many details. In particular, several symmetries about offsets of the index may be exploited. At the first stage of Fig 8.12, all the multipliers are equal to $W_N{}^0=1$, so no actual multiplications need take place; all the butterfly inputs are adjacent elements of the input array x(t). At the second stage, all the multipliers are either $W_N{}^0$ or integral powers of $W_N{}^{N/4}$ and the butterfly inputs are two samples apart, and so forth. Note that the coefficients are indexed in ascending order. These are normally calculated ahead of time and stored in a table. Another way is to use a *recursion formula* to generate them on the fly, but this is discouraged because of numerical-accuracy affects that destroy the efficiency of the technique.

All those multiplications and additions take their toll on the numerical accuracy of our final result. Quantization noise is multiplied and added as well, and at the output of a DFT, the noise power grows by N times. In an FFT calculation, the situation is roughly the same; however, the requirement to avoid overflow at intermediate stages may force us to scale the data, the coefficients, or both. This further reduces the dynamic range of any FFT. Results have been offered indicating noise increases in the vicinity of 12N. In addition, the quantization-noise contribution of the coefficients increases in inverse proportion to p, the number of bits used to represent them. This, in turn, means that the noise increase with respect to N is slow.

In FFT-based noise-reduction systems, we perform some modification of the frequency-domain data, such as zeroing bins not meeting a pre-defined amplitude threshold. Then we transform the modified data back to the time domain. The duality of the Fourier transform and its inverse can be shown in the flow diagram of a FFT^{-1} as in **Fig 8.15**. This diagram was produced from

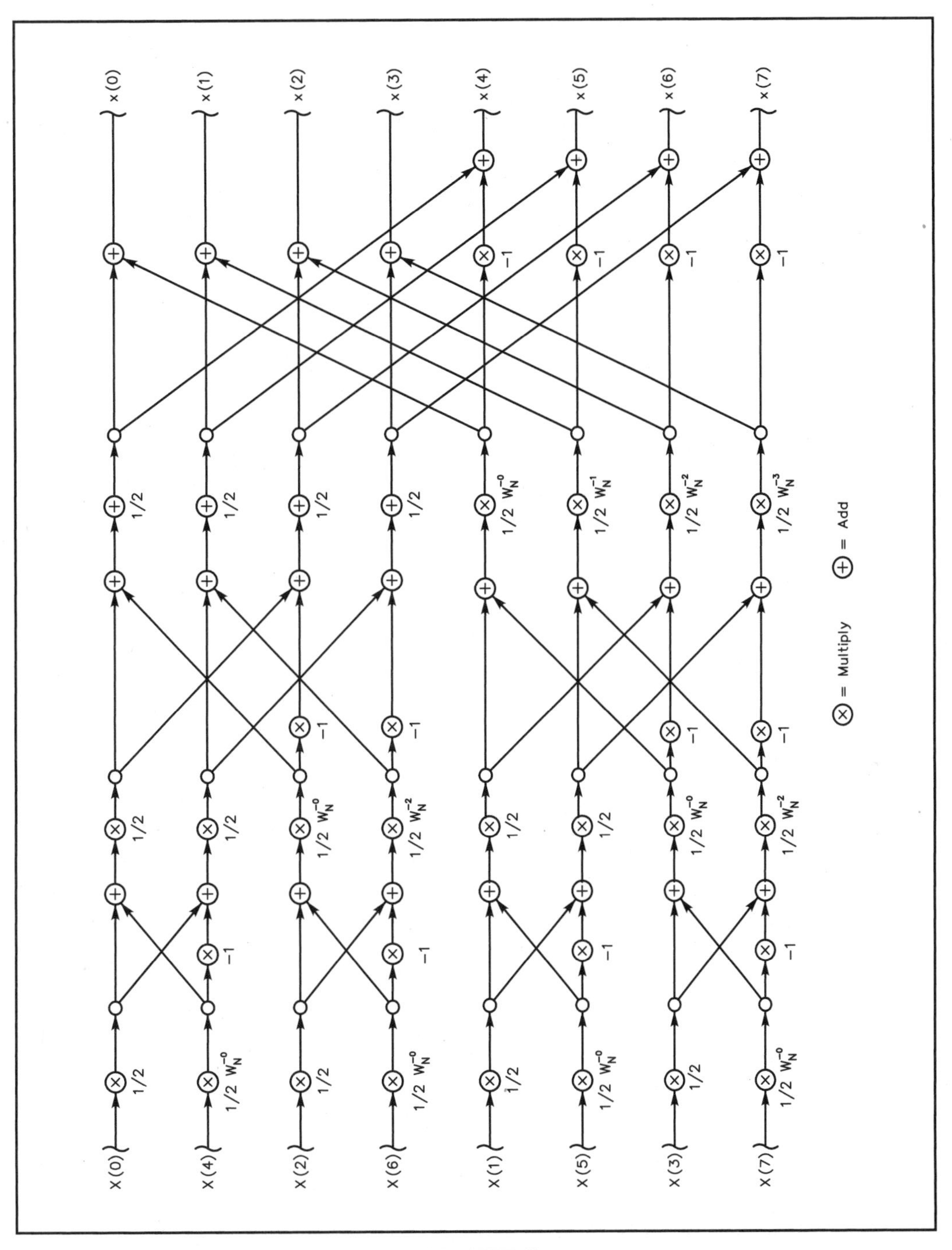

Fig 8.15—Block diagram of an inverse FFT (FFT-1).

Fig 8.12 by simply substituting $\frac{1}{2}W_N^{-kn}$ for W_N^{kn} at each stage and, of course, using X_k as the input to obtain x_t as the output. Alternatively, we may compute the FFT^{-1} by using an FFT flow diagram, swapping the inputs and outputs and reversing the direction of signal flow. It is important to note that this is a consequence of that fact that we can rearrange the nodes of the flow diagrams however we want, so long as we do not alter the result. The transforms work just as well in reverse as they do in the forward direction.

Damn-Fast Fourier Transforms

When it is necessary to compute Fourier transforms on a sample-by-sample basis, or where frequency resolution must be non-uniform across the sampling bandwidth, even traditional FFTs may be too computationally intensive for the processing horsepower available. A class of algorithms that computes the next transform output very rapidly—based solely on current transform output and the next input sample—is presented here. A method is included for controlling its inherent divergence problem by brute force.

The derivation begins by looking at how the Fourier transform results change for each bin at each sample time.[60] Say we start with some discrete Fourier transform output bins X_k at sample time r. Then we compute the DFT for the next sample time r+1 and examine the sequences to see what has changed. For r=0, each DFT sequence expands to:

$$\begin{aligned} X_{0_k} &= W_N^{0k} x_0 + W_N^{1k} x_1 + W_N^{2k} x_2 + \cdots + W_N^{(N-1)k} x_{N-1} \\ X_{1_k} &= W_N^{0k} x_1 + W_N^{1k} x_2 + W_N^{2k} x_3 + \cdots + W_N^{(N-1)k} x_N \end{aligned} \tag{117}$$

What is evident is that each input sample x_t that was multiplied by W_N^{nk} in the summation for X_{0k} is now multiplied by $W_N^{(n-1)k}$ in the summation for X_{1k}. The *ratio* of the two sequences is nearly:

$$\begin{aligned} \frac{X_{1_k}}{X_{0_k}} &= \frac{W_N^{(n-1)k}}{W_N^{nk}} \\ &= W_N^{-k} \end{aligned} \tag{118}$$

We still have two terms hanging out of the relationship, namely the first and the last:

$$W_N^{0k} x_0 = x_0 \text{ and } W_N^{(N-1)k} x_N \tag{119}$$

that have not been accounted for in the ratio. If we first subtract x_0 from X_{0k} before taking the ratio, then add the new term $W_N^{(N-1)k} x_N$ after, we have the correct result:

$$X_{1_k} = W_N^{-k}\left[X_{0_k} - x_0\right] + W_N^{(N-1)k} x_N \tag{120}$$

Now this may be simplified a little, since:

$$\begin{aligned} W_N^{(N-1)k} &= e^{\frac{-2\pi j(N-1)k}{N}} \\ &= e^{\frac{-2\pi jNk}{N}} \cdot e^{\frac{2\pi jk}{N}} \\ &= W_N^{-k} \end{aligned} \tag{121}$$

and substituting:

$$X_{1_k} = W_N^{-k}\left[X_{0_k} - x_0 + x_N\right] \tag{122}$$

This is the damn-fast Fourier transform (DFFT). It means: For N values of k, we can compute the new DFT from the old with N complex multiplications and 2N complex additions, or a computational burden proportional to N. If we begin with $X_{0k}=0$ and take the first N value of $x_N=0$, we can start the thing rolling. It saves computation over the FFT by a factor of:

$$\frac{N\log_2 N}{2N} = \frac{\log_2 N}{2} \tag{123}$$

which for large values of N is very significant indeed. For example, if N=1024, the improvement is by a factor of five. Over the direct-form DFT, it is a factor of N^2/N faster. But there is a catch: An error term will grow in the output because the truncation and rounding noise discussed previously is cumulative. The error will continue to grow unless we do something about it.

The simplest way to handle the situation is to compute two DFFTs for all the output bins k, resetting every other block of N input samples to zero. In other words, one DFFT begins at some time with an input buffer that has been zeroed; the other continues to operate on the continuous stream of real input samples. As sample-taking continues, DFFT output is taken from the second calculation. As the buffer of the first DFFT gradually fills with real samples, the block of zeroes it originally held disappears. At this point, each DFFT produces the same result except for the greater error in the second DFFT because of truncation and rounding effects. Output is then taken from the first DFFT and the buffer of the second is zeroed; the calculations continue for another N iterations, at which time the exchange and reset are again done, and so forth, continually. This places an upper bound on the cumulative error to that associated with 2N iterations and increases the computational burden by a factor of two. Now the savings over the FFT is only:

$$\frac{\log_2 N}{4} \tag{124}$$

which for N>16 still represents an improvement. DFFT output quantization noise is at least twice that of the DFT.

Frequency resolution of DFFTs is controlled by the block length, N, used in the calculations, just as in DFTs or FFTs. Resolution may be set differently, though, for each bin; further, not all bins need be computed to compute any particular bin, unlike the Cooley-Tukey FFT.

Is there an inverse DFFT? Well, because inverse Fourier transforms map into the time domain, it is simple enough to just compute the next output sample rather than the next N output samples. The easiest output term to compute is x_0, since all coefficients are $W_N^0=1$. The output is then just:

$$x_0 = \frac{1}{N}\sum_{k=0}^{N-1} X_k \tag{125}$$

and only one multiplication is involved.

Digital Transceiver Architectures

Like it or not, the communications world is going digital in every way it can. Why are we doing it? Well, previous chapters have outlined some of the reasons. Digital algorithms for filtering, modulation and demodulation, characterization of speech signals and a vast variety of other functions tend to repeat almost exactly from unit to unit. Furthermore, DSP implementations of radio transceivers usually eliminate a great deal of the hardware that traditional designs normally require. Error detection and correction is possible with digital transmission formats to a much greater extent than with analog. Finally, flexibility is retained in large measure by DSP radios. General-purpose DSPs may be coupled with data conversion hardware to produce almost any system you can think of.

The methods detailed in this book may be combined in many different ways to produce useful communications circuits. In this chapter, we shall examine several common implementations of those concepts to achieve the goal of getting information from one point to another via radio. Here, specific modes are not discussed; the material focuses on ways of arranging signal processing elements—both digital and analog—to produce working systems. Those elements are selected and discussed in terms of their impact on performance of radio transceivers overall.

General Considerations

We may define the prime goal of communications as getting an exact copy of the information that was transmitted to the receiver, without error. The enemies are noise and interference. We have looked at a few DSP approaches for engaging and defeating those enemies; but so far, we have not taken the battle all the way to the antenna.

The aspirations of a digital receiver designer are to digitize the cleanest copy of a received signal as best he or she can, then to demodulate it to extract the original information. It follows that if a signal cannot be received cleanly prior to digitization, no amount of processing could bring back the intelligence it contains. So, no matter what the transmission medium, the necessity of overcoming the limitations imposed by time and space separation cannot be avoided. Still, DSP performs some astounding magic that even ten years ago might have been considered quite implausible.

Transmitting signals cleanly is not very much of a problem. While there may be things the sender can do to help the receiver, the receiver has the far more difficult job of the two. We shall therefore concentrate on receiver design in this chapter. We begin with a peek at what DSP can do at the end of a radio-transmission chain that involves a traditional, analog receiver. As we walk the digitization point closer to the antenna, we finally approach an ideal digital receiver: It is called *direct digital conversion (DDC)*. Emphasis is placed on design trade-offs that come about because of the cost and complexity of hardware available at the time of this writing.

Mother Nature Sets the Rules

Superheterodyne receivers have been around for a while; until and unless DSP hardware can achieve sufficient speed and accuracy to digitize signals straight from the antenna, we will continue to use them. The main advantage of a superhet is that signals are converted to a fixed IF where most of the gain and selectivity is obtained. To avoid spurious responses, multiple frequency conversions are common. Recognize, though, that minimizing the number of conversions also diminishes the number of oscillators and, therefore, the number of possible unwanted internal signals or "birdies."

At some point in a digital receiver, signals get digitized and then some filtering and other processing is performed on them. We want this point to be as close to the antenna as possible, so one of the first tasks in designing a DSP receiver is to study what frequencies, bandwidths and dynamic ranges are possible in data-conversion hardware before choosing an IF. The first trade-off is usually between high-speed ADCs and the costly analog filters and amplifiers they would replace. That decision is driven mainly by cost, although issues of current consumption and processing power definitely come into play.

The final compromise also depends on the performance levels we expect to achieve. For example, many excellent ADCs are quite capable of digitizing signals directly from the antenna; their sampling rates are fast enough, but their dynamic ranges are generally not good enough at the time of this writing. HF receivers, in particular, must handle a tremendous range of input signal levels without flinching. So before making this first decision about a receiver's conversion scheme, we must think about dynamic range: What is it and how much do we want?

The Many Faces of Receiver Dynamic Range

Sensitivity must be specified with reference to some bandwidth of interest, because we are trying to copy a narrow-band signal in the presence of noise, which exists at every frequency. In that specified bandwidth, a signal received at the antenna terminal has a certain SNR. A receiver designer fights to preserve that SNR throughout the receiver. Electronic circuits of any kind always introduce some additional noise, though. The ratio of a circuit's output SNR to its input SNR is referred to as its *noise figure*.

Originally explained in full by Albert Einstein (1879-1955),[61] thermal or Brownian motion of atoms and free electrons in any conductor produces an available noise power in watts of:

$$P_{noise} = kTB \tag{126}$$

where k is Boltzmann's constant (1.38×10^{-23}), T is the absolute temperature in kelvins and B is the bandwidth in hertz. Crank some typical figures into Eq 126 and you will find that at room temperature (293 K) and in a bandwidth of 3 kHz, this power is about –139 dBm or 12.1 attowatts (12.1×10^{-18} W). This quantity represents the minimum discernable signal (MDS) in a perfect receiver at typical voice bandwidths. Note that as the temperature decreases, sensitivity increases linearly; a receiver operating in a liquid-nitrogen bath is a real gem. Atmospheric and even cosmic noise usually exceed this theoretical limit by a long way, though.

The best receivers today exhibit noise figures around 7 dB or so. When noise power equals signal power, output SNR is 0 dB and the input signal level is thus:

$$P_{in_{0\,dB}} = -139+7 = -132\,\text{dBm} \tag{127}$$

We may take this as our MDS goal and define it as the lower limit of our receiver's useful dynamic range. –132 dBm is about 0.06 μV into 50 ohms. Because of the manifold ways receivers degrade at high input levels, it is not so simple to define the upper limit of dynamic range. In fact, we shall define it in several different ways.

Normally we think of overload phenomena as involving strong, off-channel signals. It is also possible to overload a receiver with a strong, on-channel signal. For most modern receivers, that level would be so high that radio communications would not be necessary: You could just open a window and shout!

Large-signal performance is typically characterized[62] by measuring the following effects:

1. *third-order intermodulation distortion (IMD_3),*
2. *second-order IMD (IMD_2),*
3. "*blocking*" or *desensitization*, and

4. *in-band IMD.*

Let us briefly examine methods for defining and measuring each of these.

IMD Dynamic Range and Intercept Point

To measure *IMD dynamic range*, inject two off-channel signals of equal amplitude and measure the degradation in receiver performance. That degradation comes in the form of an undesired, on-channel signal produced by mixing of the off-channel signals. Increase the off-channel signals' amplitudes until the on-channel signal power equals the noise power (the MDS). Define the IMD dynamic range as the ratio of one of the off-channel signal's power to the MDS.

In the ARRL method for measuring IMD_3, one interfering signal is placed 20 kHz from the center of the channel and the other 40 kHz away. The *third-order intercept point (IP_3)* is calculated by assuming the receiver distortion obeys a perfect *cube law*. That means for every dB of increase in the off-channel signals, the IMD_3 product increases 3 dB. The difference between the IMD_3 product and the interference, therefore, increases by 2 dB for every dB increase in the off-channel signals. IP_3 is extrapolated from the measurement by adding half the IMD dynamic range to the interference level attained in the measurement, or:

$$IP_3 = \left(\frac{IMD_3\ DR}{2} \right) + P_{QRM} \tag{128}$$

That is supposed to be the level where the IMD_3 product would be equal in level to the interference. Were we to actually inject QRM at this level, though, we might find a real IP_3 much higher; receivers seldom behave perfect cube laws as they are predicted to do. This normalized procedure is a good basis for comparison, though, and it has become standard.

In the IMD_2 test, inject two non-harmonically related signals and look for an undesired product at either the sum or difference of the two signals. IMD_2 DR is measured the same way as above and IP_2 is obtained by assuming the receiver obeys a perfect *square law*: For every dB of increase in the interference, the IMD_2 product increases by 2 dB and the difference increases by 1 dB:

$$IP_2 = (IMD_2\ DR) + P_{QRM} \tag{129}$$

"But," you say, "how can a receiver obey two apparently conflicting laws at the same time?!" The answer is that in the IMD_2 case, the two fundamentals are mixing directly; whereas in the IMD_3 case, the fundamental of one frequency mixes with the internally generated second harmonic of the other. Note that when you add two fundamental frequencies, the result is always greater than twice one of the frequencies. For that reason, IMD_2 performance may be improved by using *half-octave band-pass filters* ahead of the mixer. Such filters, when switched or tuned as the receiver changes frequency, always attenuate at least one of the interfering signals, reducing the deleterious effects.

Blocking Dynamic Range

In this measurement, inject a single off-channel signal and look for some degradation of on-channel performance. In the ARRL method, the power of a single on-channel input signal is monitored. The QRM, which is placed 20 kHz away, is increased until the desired signal's output power (at the demodulator output) either increases or decreases by 1 dB.

A decrease is supposed to indicate that some stage or other is saturating, hence the term blocking. An increase results in a noise-limited measurement. *Blocking dynamic range (BDR)* is calculated as the ratio of the QRM at that point to the MDS.

In reality, saturation seldom occurs in modern receivers before the noise takes over. This noise is usually the result of *reciprocal mixing*, wherein the QRM mixes with noise sidebands on the LO to produce in-band noise.[63] Hence, BDR is often a measure of the phase-noise performance of a receiver's synthesizer more than anything else.

In-Band IMD

This is a measure of distortion produced by a receiver when the only signals present are inside the desired passband. IMD product levels are examined with respect to one of two tones presented. It is perhaps surprising to note that many receivers produce as much in-band IMD_3 as the transmitters they are listening to!

With all that under your belt, let us now take a look at putting some DSP horsepower at the output of a standard, analog receiver and see what is possible without changing anything about the way the receiver is designed.

Superheterodyne With Baseband DSP

It is common these days to apply DSP techniques at audio or baseband, especially by using outboard processing units with older, analog receivers. A drawback to this approach is that while a receiver's selectivity may be improved this way, special gain-control algorithms must be employed unless you turn off the receiver's AGC. To see why this is so, let us look at a typical arrangement, shown in **Fig 9.1**. The receiver's bandwidth is 3 kHz and we wish to use an outboard DSP unit to implement an RTTY filter with BW = 500 Hz. It follows that some of the signals we digitize are undesired and this raises a problem. When the desired signal is strong relative to the undesired, everything is fine; however, when a strong undesired signal appears within the receiver's bandwidth but outside our DSP filter's bandwidth, the receiver's AGC acts on the combination, reducing the level of our desired signal as well as that of the undesired signal. We may elect to solve this in several ways; each

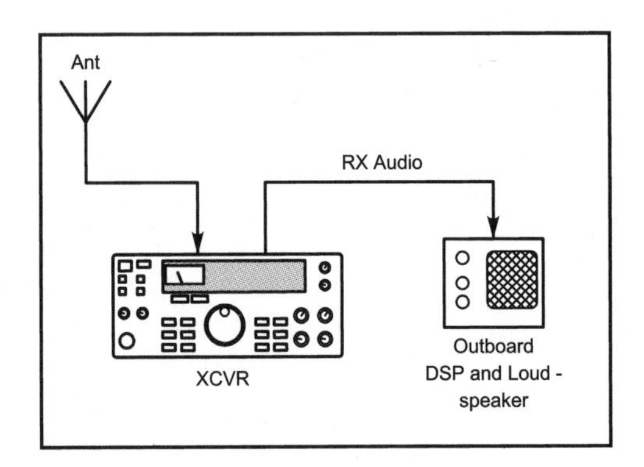

Fig 9.1—Typical use of an outboard, baseband signal processor.

involves digitally adjusting the gain applied to the desired signal to keep its level constant. Keeping the desired signal's level constant is the goal of any AGC system.

A block diagram of one such *digital AGC system* is shown in **Fig 9.2**. It consists of a gain-control block (multiplier) and a power-ratio detector. The detector computes the ratio of the sum of the undesired signal's peak amplitude, m, and the filtered signal's amplitude, n, to n:

$$k = \frac{m+n}{n} \tag{130}$$

where k is the factor by which the filtered output must be digitally boosted to keep its peak level constant. This detector includes a fast-attack, slow-decay filter. The decay rate is chosen to match that of the receiver's analog AGC. Analog AGCs usually have a decay that—when plotted as dB *vs.* seconds—looks fairly close to a straight line. Such exponential decay is achieved in DSP by multiplying the stored detector value by a constant near unity at each sample time. When the ratio k suddenly jumps upward, the stored value is updated immediately to get a fast attack.

The gain-controlled stage is simply a multiplier; its inputs are k and the filtered signal. Note that $k \geq 1$ always; hence, the multiplication is not the simple fractional type described in previous chapters. We may now have a need to extend our fixed-point math to values greater than one. This is tedious but not too difficult. We just handle the integer and fractional parts separately. Additions and subtractions are straightforward, but we have to multiply k by a fractional decay factor, δ, then multiply k by another fraction—the filtered signal—at each sample time. Separating the integer and fractional parts by a radix point, we adopt the notation k = (a.b) where:

$$a \in \mathcal{I},\ b \in \mathcal{F} \tag{131}$$

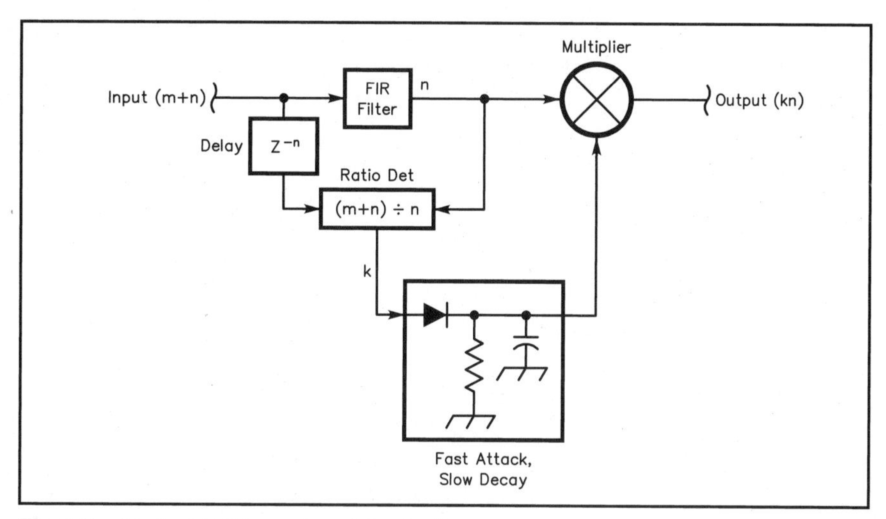

Fig 9.2—Digital AGC system block diagram.

meaning that we treat a as an integer, and b as a fraction. Numbers a and b are ordinarily represented in binary. A number whose absolute value is less than unity has a zero integer part: = (0.d). The result of the multiplication (a.b)(0.d) = e.f is:

$$(a.b)(0.d) = \left[\Im(ad)+\Im(bd)\right] . \left[\mathcal{F}(ad)+\mathcal{F}(bd)\right] \qquad (132)$$

requiring four real multiplications, just as in complex math.

A delay is inserted in the path of input signal m to compensate the delay through the DSP filter. Scaling might be necessary to prevent overflow in the gain-controlled stage. Special attention must be paid to what happens during the attack time. Some receivers exhibit *AGC overshoot*, which may cause spikes on incoming signals, resulting in rapid gain excursions. A good approach seems to be to allow gain adjustment in proportion to the attack time of input signal m, but only if m persists at that level for several milliseconds, to avoid triggering on noise pulses.

In practice, baseband DSP filtering may be limited by *in-band IMD* and synthesizer phase-noise effects that plague the analog transceiver itself. These cause unwanted signals to appear in the passband, masking the desired signal. With a perfect receiver, performance is limited by the available SNR and SFDR of the ADC in use and by the phase noise of its clock. Noise-reduction algorithms may be very effective, though, even in the face of these margins. As we move the digitization point closer to the antenna, converter noise and phase-noise issues become more critical; other factors actually aid in the resolution of some of the problems outlined above as we go to IF-DSP.

IF-DSP at a Low IF

The primary reason for wanting to digitize signals closer to the antenna is to eliminate expensive filters and other hardware whose functions can be performed in DSP. By going to a low IF, we can get rid of balanced modulators and multiple crystal or mechanical filters; demodulation, squelch, and digital AGC duties are also done in software. Many things judged quite difficult or impossible in the analog world may be included, as well.

To do it in a receiver, we apply harmonic sampling and a fast sigma-delta ADC having 16-bit resolution or better at an IF just above the audio range. This IF is selected to be comparable with the ultimate BW of the *roofing filters* used in the receiver's front end. Recall that in harmonic sampling, the sampling frequency may be as low as the IF minus half its BW. An IF BW of 15 kHz, for example, requires a sampling frequency of at least 30 kHz. The center IF itself may be almost anything greater than 7.5 kHz at this minimum sampling frequency. We ought to consider what *image rejection* we are going to get based on such a low IF, though. Roofing and other analog filters will determine it by their attenuation at an offset from center equal to twice the IF. If we intend to use the same IF in transmit mode, the 2nd LO will appear at an offset equal to the IF. Quite a few poles of filtering in the analog sections are still required around this arrangement. See **Fig 9.3**.

From the antenna, signals are low-pass filtered to remove 1st-mixing image responses and to eliminate LO leakage. Then, they are mixed to a VHF 1st IF to dodge as many spurious responses as possible. A VHF 1st IF may be selected above twice or even three times the highest RF to get away from 2nd- and 3rd-order mixing products. Six to eight poles of crystal BPF may be used in the strip, with several gain-controlled stages interspersed. A traditional IF analog AGC is employed. In any receiver design, it is best to distribute gain and loss evenly to avoid degradation of the SNR under reduced-gain conditions. We would like SNR to continue increasing as the input signal increases, as far as possible. Gain reduction, therefore, is usually made to occur at the stages closest to the antenna first, followed by subsequent stages.

First-IF signals are converted directly to the low IF, then amplified and possibly filtered further. Enough amplification is needed to raise the 2nd-IF signal to within about 10 dB of the maximum ADC input level. That maximizes the SNR available from the ADC and the dynamic range available for digital AGC operation, as described in the Baseband section above. A 10-dB margin leaves the *headroom* needed to accommodate analog AGC overshoot and noise spikes. Overload of the ADC is catastrophic and must never be allowed to occur. IF-DSP architectures that include analog AGC usually must also have digital AGC loops. Note that it is possible to build an analog front end that has the required

Fig 9.3—IF-DSP receiver block diagram.

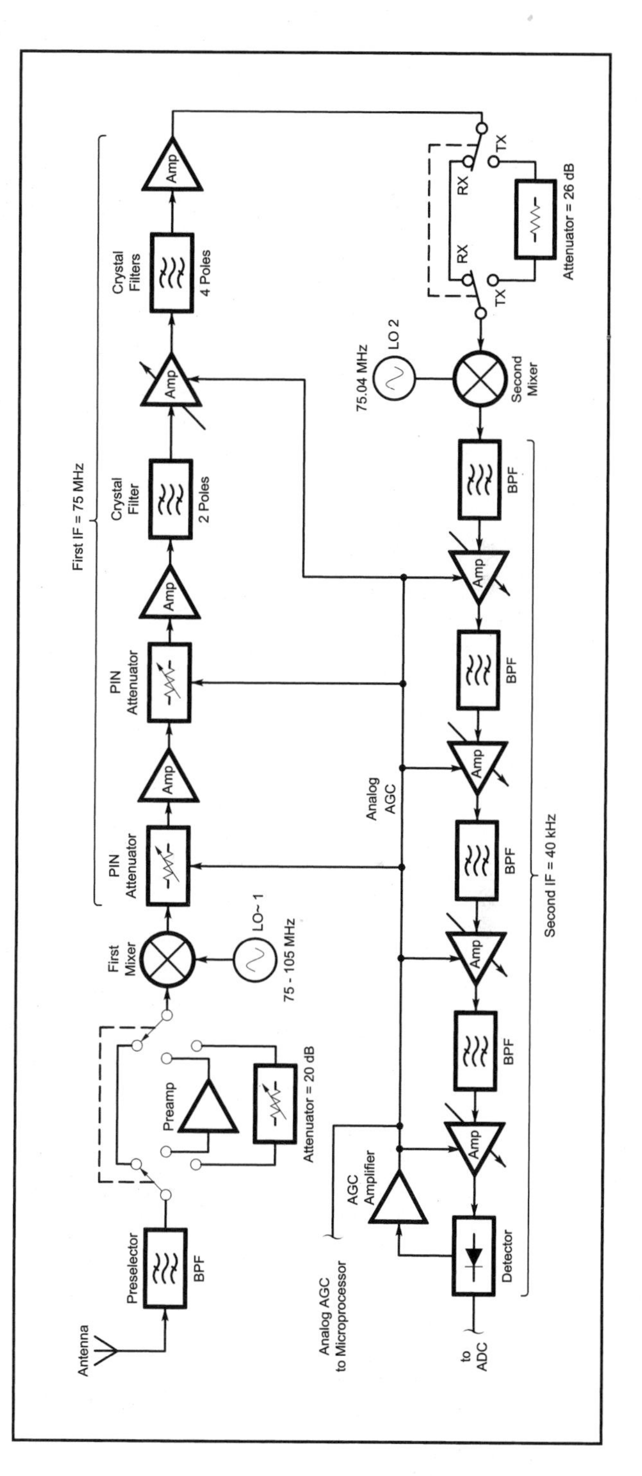

dynamic range (115 dB) and be rid of analog AGC, but this would entail some expensive hardware and tighter tolerances. Analog AGC makes it easier to keep the front end linear over the range of input signals expected. These days, receivers may be called on to handle input signals as large as one watt.

One way to handle the situation is to employ a gain-compensation scheme such as that described in the outboard-processing section above. With that, though, we may not have quite the best solution, since when the desired signal *decreases* in amplitude, the system cannot tell if it was because of QRM-caused analog gain reduction, or because the other station faded—or just stopped transmitting! So we arrange for the DSP to monitor the analog AGC voltage to find out what it is doing. Knowledge of the receiver's AGC voltage-*vs*-gain curve leads to a new algorithm: When the analog gain is changing rapidly, rapid changes in gain-control variable k are allowed, as in the baseband case. When time constants are set correctly, we do not even have to know the exact amount of analog gain reduction. If we want the digital AGC's threshold to reside above the receiver's noise floor, though, we *do* have to know the gain reduction accurately.

Traditionally, HF receivers have been designed with analog AGCs that have thresholds or "knees" around 3 μV. This means that signals below that threshold are not gain-controlled; when input signals are low, the receiver gets quiet. Note that this parameter is unrelated to the SNR as determined by the noise floor of the receiver. Digital AGCs may be designed with any threshold—it may even be made variable via a front-panel control. The effect is much the same as that produced by an IF-gain control. A knee-less digital AGC system may result in operator fatigue, because receiver and atmospheric noise are boosted in the absence of strong input signals.

A Variable Digital AGC Threshold

To restore the knee in our digital AGC, we must re-examine the dynamics of the system and redefine the goals in terms of actual system behavior. In the knee-less system, the goal was to hold the desired signal's peak output level, n, constant. We measured the peak level of the total digitized signal as the sum of the desired, n, and the undesired, m. Now we wish to perform gain compensation on the desired signal only when its level is above some preset threshold.

To begin the analysis, we will consider three cases: In case 1, m and n are both under the analog knee; in case 2, m and n are both over the analog knee; and in case 3, m is over the analog knee, but n is under it. We'll have to determine what amount of analog gain reduction was caused by m and n separately to make it work.

A Case History

In case 1, no analog gain reduction occurs. To make the digital AGC knee equal to the analog AGC's, we would set k=1. In case 2, since the desired and undesired signals are each large enough to actuate analog AGC by themselves, the correct value for k is almost the same as in the knee-less system: it is the ratio of the total signal to the desired as shown in Eq 130. We know m+n is nearly constant, and we are tempted to just use Eq 130. Case 3 is where the headaches start because m alone is responsible for the gain reduction, and we have to know its actual amount. The most straightforward method is to build a table of the gain reduction factor versus analog AGC voltage. As each receiver may have a slightly different characteristic, we may want to adaptively alter the table, or we may calibrate the table on a unit-by-unit basis. The goal is to formulate the three cases into a uniform system. As each of the cases is likely to occur in our receiver, and because rapid switching between states is possible, we need a case-detection algorithm which operates in "real time."

Let us try a comparison between the analog gain reduction value k from the table and the wanted-to-actual output ratio (Eq 130). Is the following true: $k_{table} \leq A_{out} / n$?

For case 1, this inequality will always be true, because $k_{table}<1$, and $n \leq A_{out}$. For case 3, it will also be true, since $n \leq A_{out}$ and $m \leq m+n$, where m is solely responsible for the gain reduction factor k_{table}. For case 2, it will be false, because $k_{table} > (m+n)/n$ and $(m+n) \geq A_{out}$. The analog gain reduction must be more than that caused by either m or n separately, each of which is strong enough to actuate analog AGC by itself. Although case 3 was the standout above, it is paired with case 1 in the final algorithm here, leaving case 2 by itself. We will use a digital gain boost factor $k=k_{table}$ in cases 1 and 3, and use Eq 130 for case 2. Note that the resulting transfer function is continuous, because the segments corresponding to the truth or falseness of $k_{table} \leq A_{out}/n$ must meet where $k_{table}=A_{out}/n$.

Making the Knee Continuously Adjustable

Although the transfer function is produced in two segments, adjusting the threshold requires only a modification of the test made in the inequality. If we apply a scaled value of k_{table} to the inequality, the knee will move downward according to the scaling factor. The test becomes: $p(k_{table}) \leq A_{out}/n$, where $p \geq 1$. The variable p has the effect of narrowing the width of cases 1 and 3 on the transfer curve. At some point, p gets very large, and the inequality is never satisfied; the knee is infinitely low, and the system is exactly equivalent to the knee-less system initially described. When p = 1, the digital knee is equal to the analog knee. For values of p between 1 and the maximum value of A_{out}/n, the knee is continuously adjustable.

For a receiver with a noise floor of 0.05 µV in a 500-Hz bandwidth, and with an analog AGC knee of 3 µV, the linear portion of the transfer function would be almost 36 dB deep. Without the adjustable threshold, that may be a bit much. While listening to a very weak signal with the volume cranked up, a strong signal may come along and blow the operator off his chair. Adjustment of the threshold allows maximum flexibility in operation. The control is actually used much as an IF gain control would be. Gain is reduced (threshold raised) while copying stronger signals to keep the noise floor quiet, and gain is increased (threshold lowered) during periods of low-level signal operation.

Another approach to analog AGC in a digital receiver involves generating the gain-control voltage using a detector in the digital portion. AGC voltage comes from a DAC controlled by the DSP. See **Fig 9.4**. The delay between detection of IF signals and the application of gain control causes the same problem as in traditional analog AGCs: The loop filter must be optimized with regard to amplitude and phase response so it can minimize overshoot. Delays encountered in DSP filters may require the use of a secondary ADC and detector, solely for AGC purposes. Some designers have even gone as far as to build separate IF strips for detection and application of analog AGC, all the way up to and including the receiver's 1st IF. Such architectures tend to rapidly become complicated. One of the other schemes detailed above should suffice for Amateur Radio applications.

Conversion plans used in IF-DSP receivers may also be used in the transmitter by simply swapping the LOs, inputs and outputs. A switching arrangement for this is shown in **Fig 9.5**. Isolation between the ports of the LO's DPDT switch must be set to the desired level of spectral purity. An example of such a

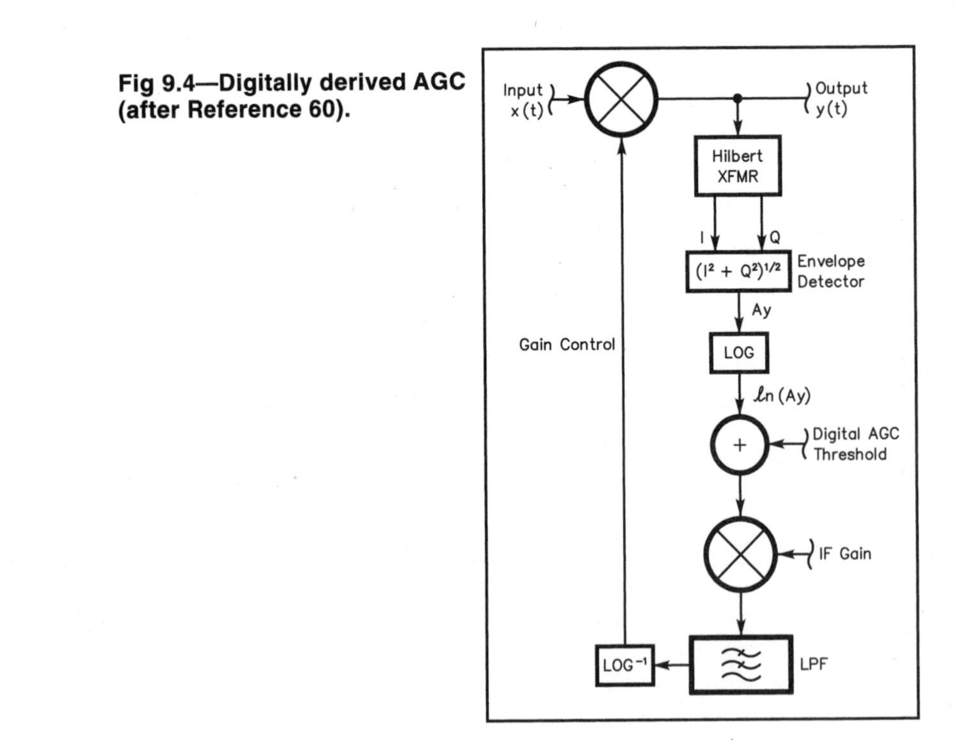

Fig 9.4—Digitally derived AGC (after Reference 60).

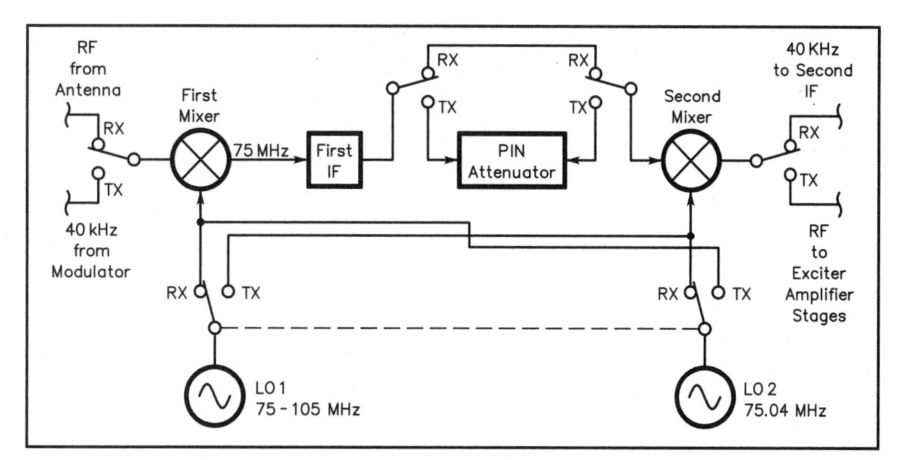

Fig 9.5—T/R switching of LOs in an IF-DSP conversion scheme.

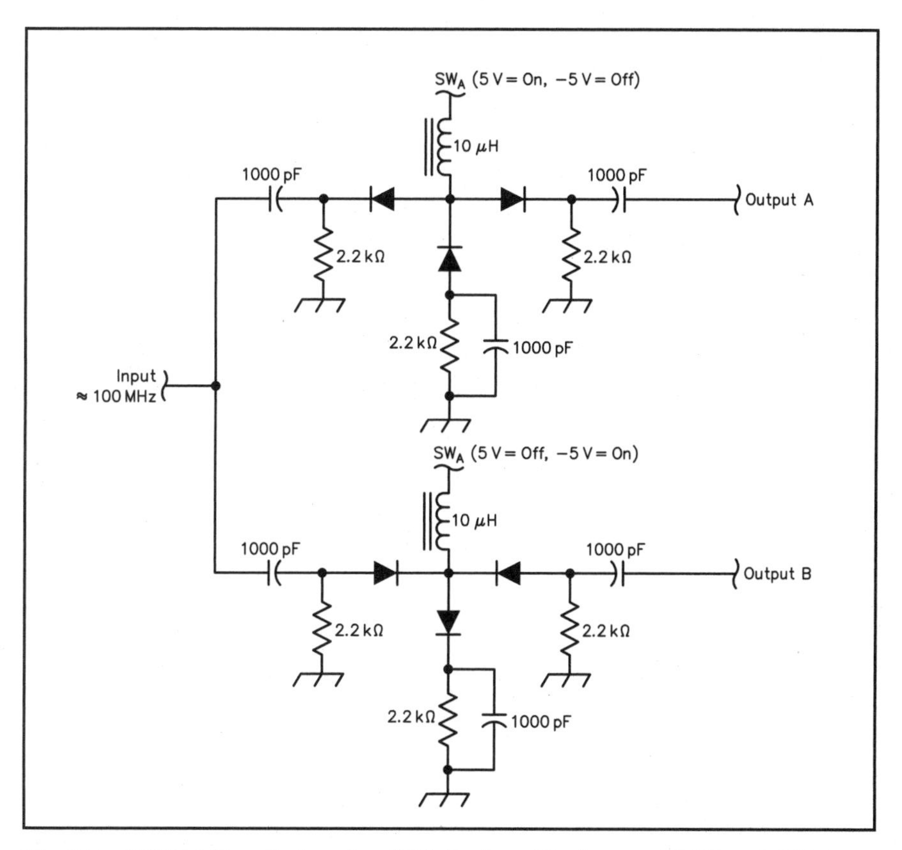

Fig 9.6—SPDT LO switch using PIN diodes. Diodes are Phillips BA682 or equivalent.

switch is shown in **Fig 9.6** using PIN diodes at VHF. Switch control voltages swing between +5 V and –5 V; when the series diodes are on, the shunt diodes are off, and *vice versa*. This particular circuit was designed for a 75-105 MHz LO and achieves better than 80 dB of isolation between the ports. Switching of the 1st mixer's input is best achieved by a relay for HF circuits, PIN diodes for VHF and above. The 2nd mixer's output may be switched using various commercial ICs, such as the Signetics NE630. Isolation in these switches is important because it reduces spurious responses in both receive and transmit modes.

Gain-controlled stages or step attenuators may have to be employed to provide for a difference in IF-strip gain between receive and transmit. To see why this might be necessary, examine the difference in gain between a receiver and a transmitter in typical service. A receiver takes as little as –132 dBm from the antenna and amplifies it to around 1 W (+30 dBm) at the loudspeaker. The power gain is:

$$GAIN_{RX} = 30 - (-132) \text{ dBm} = 162 \text{ dB} \qquad (133)$$

In a transmitter, a typical dynamic microphone might produce 5 mV rms into 600 Ω, or –44 dBm. To get to 100 W or +50 dBm, the gain is:

$$GAIN_{TX} = 50 - (-44) \text{ dBm} = 94 \text{ dB} \qquad (134)$$

The receiver has the far more difficult task, but the transmitter is still doing yeoman's duty. Considering a maximum path loss of:

$$LOSS_{PATH} = 50 - (-132) \text{ dBm} = 182 \text{ dB} \qquad (135)$$

It is a wondrously large amount of enhancement we get from our electronics, since the total power gain from the microphone on one end to the loudspeaker on the other must be:

$$GAIN_{TOTAL} = 162 + 94 = 256 \text{ dB} \qquad (136)$$

or a factor of 4×10^{25}!

Direct Digital Conversion

In the ultimate digital receiver, signals are sampled directly from the antenna without any intervening stages. In practice, though, some gain is required ahead of the ADC because of the present limitations of data-conversion technology. Quite often, one mixing stage is also included. As far as gain stages can be made with high dynamic range and good large-signal-handling capability, direct digital conversion (DDC) comes within reach. In this technique, signals are converted directly to baseband without any intermediate analog mixing stages. Refer to **Fig 9.7**. The LO is, in effect, placed very close to the desired signal and through harmonic sampling and decimation, translates it to baseband. The closeness of the LO to the signal accentuates phase-noise effects such as reciprocal mixing (discussed above) and makes short-term clock stability a major issue. Fortunately, low-noise, crystal-derived clock designs are becoming

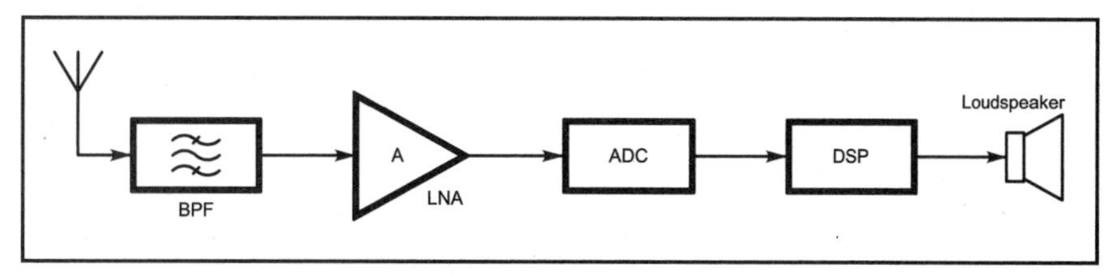

Fig 9.7—DDC receiver block diagram.

available. Even so, many DDC designs use a single mixer and a conventional LO to reduce phase-jitter effects. RMS clock jitter is usually specified in units of time (picoseconds rms), but a clock's phase-noise-*vs*-frequency characteristic tells the entire tale.

The Nyquist criterion compactly determines the sampling rate required for any given signal or group of signals. If the digitized BW is 50 kHz, the minimum sampling rate is 100 kHz, even if the signal's frequency is in the VHF range or beyond. Ancillary sample-and-hold devices may be employed in a DDC receiver to ease the BW requirements for the ADC. The digitized BW must remain within half the final sampling frequency to avoid aliasing; for this reason, interest in narrow pre-selector filters has been renewed. In the example of a 50-kHz received BW, any increase in sampling rate above 100 kHz is called *over-sampling*. Over-sampling is important because it allows for an SNR gain by spreading quantization noise over a larger BW, then filtering it, as discussed above. We are using harmonic sampling, though, so we are also *under-sampling* our signal. We can be both over-sampling and under-sampling at the same time because one is defined with respect to BW and the other by the frequencies of interest.

Frequency planning is of special concern in DDC architectures. Quite often, spurious responses appear in high-speed ADC and DAC outputs that we must plan to avoid. Those are largely responsible for establishing the SFDR of high-speed converters. Problems are also created in supposedly linear stages that generate significant harmonic content, since those harmonics show up as aliases in the digitized spectrum and may mix with other products. Careful selection of sampling frequency and IF can place these spurs where they are harmless: outside the band of interest. Over-sampling only moves us toward the goal by providing more spectrum into which spurs may harmlessly fall.

The technique known as dithering further improves SFDR by spreading the energy contained in spurs—those caused by the DNL of data converters—over greater BWs. Dithering artificially adds noise to the clock input of the ADC or DAC to achieve this spectral spreading. Typical values used are in the range of 10-30 bits peak-to-peak. Spurious reduction of over 20 dB has been attained with high-speed (40-Msample/s) converters.

Hardware for Embedded DSP Systems

What is it about a microprocessor that makes it a DSP? Well, DSPs are special because they include facilities uniquely designed for the type of calculations common in signal-processing algorithms. Almost all of them are 16-bit machines, or better, and so are very powerful even without their special facilities. DSPs may be classified primarily by their representation of numbers (fixed-point *vs.* floating-point), also by their data-path width (16-bit, 32-bit), by their programmability (general-purpose *vs* dedicated coprocessor), and their speed.

Fixed-Point DSPs

Fixed-point DSPs are generally simpler than floating-point units, so they are typically less expensive. Fixed-point processors are common in embedded systems, especially for radio. Special software instructions and separate high-speed computational units are included to accelerate the processing of common DSP calculations. Perhaps the most-used operation is the convolution sum, performed as a series of MAC instructions (see Chapter 4). Designers are interested in how many MACs per second a DSP can execute, because for anything beyond simple audio processing, only a small amount of time is available between samples for filtering and other functions.

A typical 16-bit, fixed-point DSP is shown in the block diagram of **Fig 10.1**. It employs what is called the *Harvard architecture*: It has separate program and data memory paths and also includes a pipeline for holding instructions waiting to be executed. This arrangement speeds things along because the CPU can fetch future instructions even when it is executing the current instruction or fetching data from another path. Consider how this affects an FIR filter algorithm, for example.

For each tap in the filter, the processor must multiply a constant (a filter coefficient) by a data value (a stored sample). When the processor can fetch

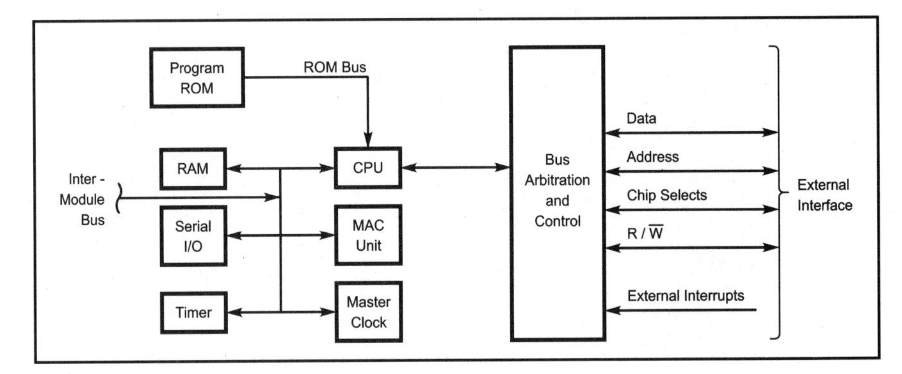

Fig 10.1—Fixed-point DSP block diagram.

both values simultaneously, an entire cycle time is saved. The subsequent addition of the product to the accumulator and the incrementing of indices for the next MAC instruction may also be executed in a single cycle. When large filters are being implemented, time savings quickly mount. Contrast this with the many cycles needed to perform the same operations in a general-purpose computer and you will see why specialized processors are so much more capable of handling sampled signals.

This business of execution speed is a large factor in the selection of a DSP for any particular use. System planning must begin by reckoning how many instructions can be executed between sample times. In a system with a 30-kHz sampling rate, only 33 µs are available, so a fixed-point DSP that can execute two million MACs per second (2 M-MACs/s) can only get 66 of these in the space between samples. For all but the simplest of systems, this is generally insufficient power for good filtering and other requirements and a separate filter *coprocessor* must be employed. This is discussed further below. DSPs are now available having over 200 M-MACs/s performance.

Many fixed-point DSPs are available that also have undedicated parallel and serial input/output (I/O) on board. Those may be very useful for embedded applications by obviating the need for other hardware. Processors embedded in radios have traditionally been shut off during times when no user input is present, stopping their clocks. That is done to eliminate the digital-circuit noise that otherwise would be difficult to remove. With a DSP in critical signal paths, this luxury is not possible. Careful attention to shielding, grounding and bypassing must therefore be paid. A DSP and associated support components humming along at 25 MHz (or more) tend to generate lots of noise and discrete spectral elements. They also tend to draw significant current, although dissipations in the 1-watt range are typical; for base-station equipment, that is not usually a big concern.

Table 10.1
Popular Fixed and Floating-Point DSPs

Part Number	*Manufacturer*	*# of bits*	*Fixed/Floating*
TMS320Cxx	Texas Instr	16	Fixed
DSP320Cxx	Microchip	16	"
DSP16	ATT	16	"
ADSP21xx	Analog Dev	16	"
MC68HC16	Motorola	16	"
MC5600x	Motorola	24	"
MB862xx	Fujitsu	24	Floating
MC9600x	Motorola	32	"
DSP32x	ATT	32	"
TMS320Cxx	Texas Instr	32	"
ADSP 21xxx	Analog Dev	32	"

Fixed-point math brings with it a limitation on the range of numbers that can be represented, notwithstanding the extended integer/fractional representation demonstrated above. This limitation may form an obstacle to achieving the highest possible dynamic range. For this reason, floating-point DSPs are also widely available for use where greater boundaries must be set on the range of numbers handled. **Table 10.1** shows a listing of popular fixed-point DSPs, along with their floating-point cousins. Manufacturers supply evaluation boards, some of which include data converters and other support circuitry. Control software that runs on a desk-top computer is available for downloading *object code*—the DSP instructions that make up the program—as well as for debugging by use of tools such as break-points and register dumps.

Floating-Point DSPs

Representation of numbers is a critical decision to be made early in the system design process. A decision to use a floating-point DSP, at generally higher cost than fixed-point, is usually made either to remove dynamic-range barriers or to grant greater flexibility to algorithms that require scaling of data and coefficients, such as the FFT algorithms discussed above. We saw that each floating-point number requires two storage locations: one for the mantissa and one for the exponent. One would expect the processing of these numbers to be slowed by having to handle twice the data, but floating-point architectures may be devised in such a way as to minimize or even eliminate this apparent handicap.

Multiplying two floating-point numbers involves multiplying the mantissas, then adding the exponents and any carry (or borrow) from the multiplication. Since multiplications generally require more time than additions, summing the exponents does not really slow the machine very much. Adding two floating-point numbers, though, requires the addition of the mantissas and a possible adjustment to the exponent, and this is always a bit slower than can be done on fixed-point numbers. With an optimized MAC unit, even this restriction can be removed for the bulk of calculations in typical DSP applications. Other than for those points, the block diagram of a floating-point DSP does not look very different from that of the fixed-point unit in Fig 10.1.

Selecting Data Converters

Complete DSP systems almost always include data converters in the form of one or more ADCs and DACs. Selection of these devices for any particular application is made with regard to cost, bit-resolution, speed, SFDR, and digital interface. Manufacturers characterize devices on these bases and obviously, we must choose them so they will handle the highest sampling rate at our analog interface. In general, bit-resolution and speed determine SFDR. Dual 16-bit ADCs and DACs are now very common because they are used in stereo compact-disc (CD) recorders and playback units at a sampling rate of 44.1 kHz. Note that 44.1 kilosamples/s of two channels in a stereo system is equal to (2)·(44,100)·(16) 1.41 megabits per second. 20- and even 24-bit devices are catching on. That is a lot of data and the bit-resolution of data converters is most often chosen to match that of the DSP, although there may be advantages in having slightly more bit-resolution in the DSP to mitigate round-off errors, as noted in Chapter 4.

We noted before that over-sampling of input signals brings significant advantages for the DSP designer. For this reason, sigma-delta ADCs are the "top of the crop" for use in IF-DSP and DDC receivers. See **Fig 10.2** for the block diagram of a sigma-delta ADC.

As sampling frequencies increase, over-sampling becomes more difficult to achieve. Engineers working in cellular radio and similar technologies deal with much wider bandwidths than most of those found in Amateur Radio, and so they must grapple with reduced dynamic range; fortunately, they also require less. 12- to 16-bit ADCs at speeds exceeding 100 MHz are available. Viable DDC designs are finding their ways into many commercial services worldwide.

Converters must interface with DSPs through a high-speed digital connection of some kind. Parallel transfer—all 16 bits at once, for example—is more

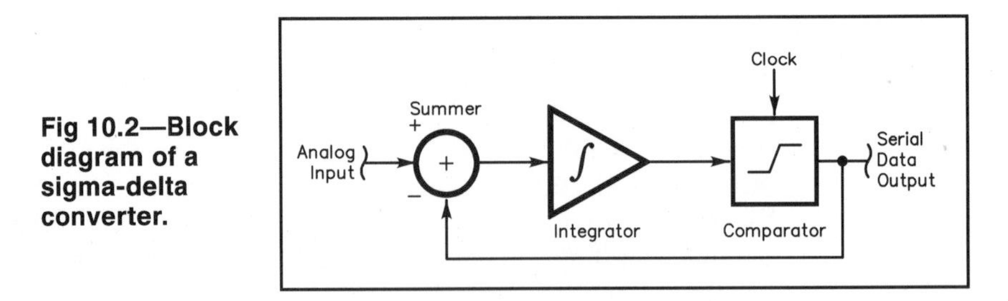

Fig 10.2—Block diagram of a sigma-delta converter.

common among DACs than ADCs. High-speed, three-line serial interfaces are popular among converter manufacturers and several standards have evolved. Some of these are compatible with one another. Bearing in mind the amount of data being transferred, realize that these serial links may run at very high clock speeds. ADC/DAC evaluation boards may be connected to DSP evaluation boards to form a prototype DSP system. Some data converters are listed in **Table 10.2**.

Extra Processing Power: DSP Coprocessors

Quite often, a single, general-purpose DSP by itself is insufficient to handle the computational load in a project. This may be determined early in the system design by evaluating the number of MACs required by filters and other algorithms. Several solutions present themselves: adding one or more general-purpose DSPs, adding specialized coprocessor chips, or designing a custom coprocessor using programmable-logic chips.

More than one general-purpose DSP may be used to augment net data capacity. The trend these days, though, is to use dedicated coprocessor chips that are optimized for the function they are to perform. This is especially true of FFT and other operations that do not lend themselves well to the MAC procedures for which general-purpose chips are optimized. Whatever the algorithm, it seems that multiplication of numbers takes the most time, so a coprocessor that incorporates a fast-multiplication algorithm is to be desired. A lot of effort has gone into fast multipliers since the 1980s and for the IF-DSP or DDC

Table 10.2
Data Conversion Chips

Part Number	*Manufacturer*	*# bits*	*Speed*	*ADC/DAC*
HI1171	Harris/Intersil	8	40 Ms/s	ADC
HI1276	"	8	500 Ms/s	"
AD7722	Analog Dev	16	200 ks/s	"
ADC76	Burr Brown	16	50 ks/s	"
PCM1750	"	18	44 ks/s	dual ADC
CS5322	Crystal	24	2 ks/s	ADC
BT254	Brooktree	24	30 Ms/s	"
Note: Also see Maxim, National, Sipex, Analogic				
CA3338A	Harris/Int	8	50 Ms/s	DAC
HI1171	"	8	40 Ms/s	"
HI5780	"	10	40 Ms/s	"
HI20201	"	10	160 Ms/s	"
PCM56	Burr Brown	16	93 ks/s	DAC
PCM66	"	16	44 ks/s	dual DAC
Note: See also National, Analog Dev, Maxim, etc.				

Fig 10.3—Long multiplication of two binary numbers.

```
    1011
  × 1001
 -------
    1011
   0000-
  0000--
 1011---
 -------
= 1100011
```

Table 10.3

Dedicated Coprocessor Chips

Part Number	*Manufacturer*	*# bits*	*Speed*	*Function*
HSP50016	Harris/Int	16	52 Ms/s	DSP down-conv
HSP50110	"	10	60 Ms/s	Quadr tuner
HSP50210	"	10	52 Ms/s	DSP Costas loop
HSP50306	"	6	2 Mbit/s	QPSK demod
HSP43xxx	"	10-24	var	DSP filters
510	Harris et al	16	10 Ms/s	Mult/Acc
LMA2010	Logic Dev, IDT	16	40 Ms/s	Mult/Acc
HSP4510x	Harris/Int	20-32	33 Ms/s	DDS
Various	Xylinx, Altera, Atmel, etc	8-32	>100 Ms/s	FPGAs

designer, a knowledge of how it is done may bring plentiful results.

The multiplication of two binary numbers may be decomposed into an addition of several binary numbers. We know that fast binary addition is readily achieved by relatively simple logic. Let's take a look at this, since it forms the basis for most fast multipliers. Shown in **Fig 10.3** is the long multiplication of two 4-bit binary numbers. It is performed in base two the same way as it is in base ten: First, take the least-significant digit of the lower multiplicand and multiply it by the other multiplier. Since in binary, this digit is either one or zero, the digits we write under the line are either a copy of the top multiplicand, or all zeros. Then, the next-significant digit of the lower multiplicand is used, with the result written below the first and shifted one digit to the left. This process continues until all bits of the lower multiplicand have been used. Finally, all the interim results are added. This last result is the product of the two numbers. Note that the result may contain a number of bits as high as the sum of the number of bits in both the multiplicands.

Refinements of this technique that use look-up tables and combinatorial methods yield speed increases. Field-programmable gate array (FPGA) manufacturers have worked out the details of these algorithms and routinely provide them to users. FPGAs are available now in very-high-speed versions ($f_{clk} \geq 500$ MHz) that may be used for DSP coprocessing. FPGA designs may also employ the Harvard architecture using external, *dual-port memory* to provide a register-based interface to host DSPs. Normally, one sample is passed to the coprocessor and one retrieved at each sample time. Filters exceeding 100 taps may be implemented this way, saving processing time in the host DSP for other housekeeping tasks.

Entire down-conversion and I/Q modulation subsystems have been incorporated on a single chip. These chip sets may be advantageous where FPGA-based designs either do not meet requirements or are too expensive. A sampling of ready-to-use coprocessors and DDC chips is given in **Table 10.3**. Also read some of the reference material listed at the end of this book for more information on dedicated DSP coprocessors.

DSP System Software

Assembly Language and Timing Requirements

Embedded-DSP application software is most often written in *assembly language*, the native language of the DSP in use. Instructions to be executed are arranged in order, according to the *von Neumann model*, and entered as lines in a text file, using the mnemonics provided by the DSP manufacturer. When this *source code* is ready, an *assembler* program is invoked that translates the source code into object code—the numbers that the DSP understands as instructions. The object code is then transferred to the program memory of the target system for execution.

The reason assembly language is so prevalent in embedded applications is the critical timing involved. Programs compiled in high-level languages do not always handle interrupt-driven events well (the input or output samples) and may bog down. To minimize the required hardware speed, processing of some second-line tasks such as squelch and ALC must have reduced sampling rates to fit into the whole picture. Only a part of their processing burden may be performed at each sample time. This is a form of *time-distributed processing* and is just one in the DSP designer's bag of tricks.

Someone will always think of something more for a transceiver to do and it is better to err on the side of higher speed and more memory at the start than to run out later. Even so, DSP designers must carefully evaluate all the functions included at the outset. Other shortcuts—like the assumption of only integer values by a BFO at one fourth the sampling frequency—may present themselves, but one cannot always count on it; one must plan diligently to avoid roadblocks. In addition, *unexpected things can occur* if due thought is not given to quantization and scaling effects, especially where adaptive processing is applied, no matter the representation of numbers used. DSP-chip manufacturers provide assemblers and instruction details free of charge. Their applications engineers are ordinarily ready to assist. A plethora of information is available on the Web.

Filter-Design Software

Several software packages for DSP filter design are listed at the end of this book. Many more are available. You can expect to find reasonably priced software that will design FIR and IIR filters, as well as let you perform convolution, multiplication, addition, logarithms and other calculations on numeric sequences. FIR filters usually may be designed with a choice of method (Fourier, Parks-McClellan, least-squares), length, frequency response, and ripple magnitude; they may use various window functions to achieve different shape factors and passband/stopband attenuations. Some are able to take coefficient and data quantization into account and some are not. Large filters may deviate significantly from their theoretical responses because of these effects, so if you are contemplating reasonably long filters, check into this capability. IIR filter design usually includes a choice of various analog-filter prototypes. Software packages may vary in their ability to display, print, or plot responses and write coefficient files to disk. Filter coefficients are generally part of system firmware and must be transferred from the host DSP to a filter coprocessor on demand. It must be possible to translate the filter-design software's output to a format the compiler software understands. A translation program may have to be written to accomplish this.

Longer and more-complex FIR filters may be implemented by convolving the impulse responses of several different filters. This allows the alteration of the frequency response of standard filters to include graphic or parametric equalization and IF shift. Such filtering systems are already being employed in Amateur Radio and commercial transceivers.

Other DSP Design Tools

FPGA design software is generally available from chip manufacturers. In addition, many schematic-capture and PCB-layout software vendors provide interfaces to popular FPGAs and other programmable devices. Hardware Design Language (HDL) and Very High Speed Integrated Circuit HDL (VHDL) have become popular for translating user requirements into programming code for FPGAs. Most FPGA programmers understand HDL or VHDL.

A rich variety of flow-chart software exists in both the public and private domains. It may be especially useful for time-sensitive applications in DSP.

Algorithm Design, Documentation and Testing

Many DSP practitioners know that software can be the black hole into which you throw a lot of your money, time and effort. It makes sense, therefore, to design system software in a very organized manner. Testing the performance of software is a critical phase in any microprocessor-based design; it should be equally well organized. Below is a brief discussion of general concepts that make the job easier and more rewarding.

A Modular Approach

Large tasks, such as the creation of DSP algorithms and their integration into a whole operating system, may be logically broken down into smaller tasks. Such *modular software design* carries with it a host of benefits. Among those are elimination of code duplication, possibilities for multiple uses of any one chunk of code, and the clear definition of input and output data. Software's intimate relationship with hardware quite often dictates how it is structured. Of particular interest to the DSP programmer are the ideas of *subroutines* and *interrupt handlers*.

Subroutines are blocks of code that perform one or more specific functions and that have well-defined inputs and outputs. A subroutine that calculates square roots, for example, might find use both in an AM demodulator and in a speech processor, as described separately above. Subroutines may be written for almost any function in a radio transceiver; when software is structured this way, code becomes easier to modify and maintain. Frequently, programmers arrange their code so that routines running in a reiterative loop simply make *subroutine calls* and pass or store data for processing by other subroutines. Subroutines may make calls to still other subroutines. The number of times that occurs is known as the *nesting* level.

When a subroutine is called, a microprocessor normally saves the address of the instruction immediately following the call so it can retrieve it and return to execution of the calling code after the subroutine has finished. *Volatile memory* (random-access memory or RAM) is usually reserved for this purpose: It is known as a *stack*. The stack may also be used to pass input data to the subroutine and recover output data from it. That arrangement is particularly slick when the operations of *pushing* and *pulling* (or *popping*) data to and from the stack involve little overhead. In one simple example, the calling code pushes (stores in a location on the stack) the input data for the subroutine. The subroutine call is then made and the return address is pushed onto the stack. The subroutine may then access the input data starting at the second stack entry as indexed by a *stack pointer*. The subroutine code therefore also has information about which part of higher-level code made the call.

The stack pointer is altered every time data are pushed or pulled from the stack. Output data may be stored in microprocessor registers before returning to execution of the calling code. At the end of subroutine code, the stack pointer must point to the return address. The microprocessor pulls the return address from the stack, updating the stack pointer in the process, and execution resumes with the instruction following the subroutine call.

When the nesting level gets large, the risk of *stack overflow* presents itself. That happens when data are pushed onto the stack beyond its predefined size. The stack pointer normally then points to its first address (instead of its last) and bad things happen, since normal instruction execution is disrupted. Similarly, *stack underflow* can occur when more data are pulled from the stack than are pushed onto it. Very bad things also occur in that case.

Interrupts and Their Handlers

Sometimes it is necessary for software to handle asynchronous events, such as the arrival of new input data for some algorithm. DSPs and other microprocessors normally provide for this situation by examining *interrupt signals*. Those are hardware-related signals that, when asserted, cause the processor to cease execution of the current instruction flow and divert to a special block of code. An interrupt is regarded as so important that the normal flow of instructions must be stopped and the interrupting source serviced.

That obviously involves a delay in the processing of main-line tasks. A microprocessor must store not only a return address on the stack, but also some information about register contents and machine state when the interrupt occurred; the machine can thereby return to its original state when higher-level processing resumes. Overhead associated with interrupts is therefore greater than that associated with subroutines.

Many microprocessors save a fixed set of register contents to the stack when an interrupt occurs; others allow the programmer to decide what part of the previous state is saved. The instruction set executed after receipt of an interrupt are known as an *interrupt handler* or *interrupt service routine*. Those instructions are normally pointed to by an *interrupt vector* that contains the address of the first instruction of the service routine. These days, most DSPs support many different interrupts, so there may be many interrupt vectors associated with various interrupt sources. For example, receipt of a command from an external controller via a serial port may force a *serial interrupt*. Alternatively, arrival of a new data sample may force a different interrupt.

Many microprocessors also provide for *software interrupts*. Those are handled the same way as hardware interrupts, except that a software interrupt is initiated by an instruction in the code. A software interrupt is another example of how chunks of code may be reused for different purposes.

Interrupt nesting level obviously is limited by the size of the stack, just as are subroutines. A return-from-interrupt instruction, like a return-from-subroutine instruction, must cause a microprocessor to pull the same number of bytes from the stack that were pushed onto it when the call was made.

Interrupts may be a convenient way to handle asynchronous events, but they are not always the most efficient approach. Overhead associated with stacking and retrieving the machine state may slow algorithms that are critical to performance. *Polled operation* may be a quicker alternative to interrupt operation. In that construct, software continually examines data sources to find whether new input is present, without using interrupts. As against that, polled operation quite often involves finagling of timing constraints that confront DSP designers.

Software Documentation

DSP software is often confusing and ambiguous; comments should be included for every line of code whose purpose is not self-evident. High-level languages, such as BASIC or FORTRAN may provide the means necessary for subsequent programmers to understand how some code functions. Those languages are sometimes used to provide comments to assembly-language code.

Plain English suffices well enough in most cases. A few words about a variable being modified or a function being performed go a long way toward understanding. You may say that if a piece of code does its job, understanding is not that important; but every blob of software depends on something that came before it, so it is important that code be well documented.

Subroutines and interrupt handlers should have clearly defined inputs and outputs. Giant segments of comments are common among experienced programmers. We often forget what we intended when we originally wrote code; those notes really help when it is time to revisit it!

Software Testing

Often, many different functions are performed in DSP that require close attention to their interaction; a precise test plan should be constructed that tries to anticipate unexpected effects. An example of this may occur when implementing noise reduction and auto-notch algorithms simultaneously. A matrix of available functions versus expected results should be constructed that provides a road map for a test plan. A person who is used to operating radio transceivers is quite often a useful "Guinea pig" for experimentation. He or she usually finds some aspect of system behavior that is contrary to intuition. DSP programmers cannot always see the forest for the trees while they are embroiled in their jobs!

Control System Theory

Control systems are based on the idea of feedback: Some command is issued to a system and information about system response is obtained in return. In many cases, verification of the response is immediately available: The radio is right in front of you. In other cases, the radio may be remotely located; all information about its performance must be gathered by its responses to commands obtained through what may be a questionable medium.

In fact, feedback may not be available; this situation is to be avoided wherever possible. In the case of a microprocessor-controlled transceiver, almost everything may be put under control of the processor; it makes sense, therefore, that it manage all control commands and feedback.

In the case of a digital transceiver, the range of functions under microprocessor control is immense. It is prudent to provide as much user control as possible. The challenge is to present controls in a way that is unambiguous and user-friendly. Many of us have encountered user-deadly operating systems that make Amateur Radio a chore, rather than the pleasure it should be. Let us look at some functions that may be included in a typical rig to show how DSP improves performance.

S-Meter Calibration

Most transceivers have an AGC system of some kind. It is desirable to meter received signal strength; this is usually done by using the AGC voltage to drive a visual display. Because of the nature of RF circuits, gain control is usually proportional to the anti-logarithm of the AGC voltage. Hence, S meters are typically calibrated logarithmically: Each increment of meter deflection is proportional to a fixed number of decibels.

Analog gain-controlled devices, though, do not repeat exactly from unit to unit. To get an accurate S meter, it is necessary to characterize each radio separately via a calibration routine. It is easy enough in DSP to measure an S meter's performance and to build a table of its errors.

A correction table may be stored in nonvolatile memory and used to adjust S-meter values on the fly. During initial testing, a calibration routine may be used to compare meter readings with the actual signal level from a known-accurate signal generator. After many units have been measured, a small number of S-meter correction curves is sufficient to account for all units; those boiler-plate tables are then used in production.

Frequency Calibration and Temperature Compensation

Another benefit of DSP control systems is that it is easy to determine the frequency of a received cw signal to within the accuracy of the control system's reference-clock. An accurate external reference frequency may be input to the receiver and its frequency counted by the DSP. That frequency error may be used to generate a correction voltage, which is then applied to the master frequency reference of the transceiver. The master reference may be a voltage-controlled crystal oscillator (VCXO).

The VCXO control voltage may be varied with temperature to compensate the oscillator's frequency-vs-temperature curve, thereby making it a digitally compensated crystal oscillator (DCXO). For best accuracy, each unit may be calibrated individually. The temperature sensor should be located in the oscillator compartment to achieve best tracking. Note that all frequency-determining elements are included in the calibration loop and are therefore compensated to some degree using this technique.

Notice that such a system is adaptive: It can accommodate errors that change with time or that are nonlinear. Also, sensitivity to the external reference used for calibration may be increased by initially using a wide bandwidth for acquisition, then narrowing it to improve the SNR and therefore, the accuracy of the result.

A Receiver as a Spectrum Analyzer

As shown before, various techniques are available to analyze signals in the frequency domain. With today's frequency-agile synthesizers, it is possible to turn a receiver into a spectrum analyzer.

To do it, the receiver is rapidly tuned across the band of interest; signal strength is measured at each iteration. A very fast AGC time constant is obviously required. The resolution bandwidth of measurement may be altered by selecting a different receiver bandwidth, with an attendant change in sweep speed. The resulting data may be graphically displayed; the user is given a graphical user interface (GUI) with which to visually select signals and tune the receiver. That is an advantage when looking for busy channels (during contests) or for idle channels (for rag chewing or CQs).

Systems for Remote Control

Hams who live in antenna-restricted areas must find some means of getting on the air without upsetting their neighbors. Antennas are viewed as an eyesore by an increasing number of neighborhood associations; but it is RF interference that gets us in the most trouble. Remote location and control of a transceiver solves both those problems.

Commercial radio systems must often crowd many transmitters into a small area. They are then forced to distantly locate the receivers. Some way of controlling those remote receivers and of coordinating their operation with that of corresponding transmitters is necessary.

Remote control allows many operators to share a single transceiver. Repeaters are a good example of that. Clubs or schools may share a single transceiver by coordinating access times. That is not as convenient as having the thing at your disposal 24 hours a day, but it sure cuts costs.

Remote control is covered here as a software issue because the controller is quite often a personal computer running some control code. Dedicated remote controllers are commonplace with commercial and military rigs, but PCs are frequently found in ham shacks and they make excellent, flexible control systems. Some ham rigs are designed to be controlled exclusively by PCs, eliminating many expensive front-panel components and the associated clutter. Some say those manual controls are beautiful things, though; others say they cost too much and they point out you cannot change their locations or hide them when you want to. Certainly, a PC with a keyboard, mouse and maybe even some voice-recognition software holds many possibilities, especially for the disabled.

Remote-Control Media

Several media are available for remote control of transceivers: the public switched telephone network (PSTN); the Internet; and a direct or satellite VHF, UHF or microwave radio link. It is desirable that command and control data traverse the control link simultaneously with the audio or data that constitute the desired communications information. Further, we would like the system to operate in full-duplex so that information can be passed to and from the control site at the same time. Finally, a control link must have a means of verifying that commands were properly executed; it also must minimize corruption of the desired communications information through error-detection and correction, if possible.

Starting in 1996, interest in simultaneous voice and data transmission intensified. The PSTN is inherently a full-duplex system; but without special techniques, it cannot support analog voice and modulated data waveforms at the same time. Driven by the demand for teleconferencing, numerous companies introduced digital simultaneous voice and data (DSVD) telephone modems. Those were followed closely by so-called analog simultaneous voice and data (ASVD) modems that touted improved audio quality while maintaining moderate data-transmission bandwidths.

In a remote-control system, the bandwidth required for control data is usually much less than that required to pass received or transmitted signals, such as audio. DSVD telephone modems are often optimized for the opposite situation: mostly data and just enough bandwidth to get communications-quality audio. Occupied bandwidth of digitized audio may be minimized using any of the techniques described in Chapter 6.

High-speed Internet access media such as ISDN, DSL and even spread-spectrum radio links have opened new possibilities for remote control. Data transfer for its own sake now commonly occupies the same Internet traffic flows as digitized, streaming audio and even video. Some hams have made their transceivers accessible via the Internet using popular Internet-phone programs such as Real Audio and Net Meeting. At the time of this writing, *quality-of-service* (QOS) issues were the primary factors influencing the performance of Internet remote-control systems. Operators have found that, quite often, bottlenecks in packet flow across the Web create breaks in digitized audio and delays in control commands that are very undesirable. An increasing number of ISPs are employing QOS software and hardware that mitigate those problems to a large degree.

The use of a direct radio link between the control point and the transceiver commonly overcomes problems like those found on the Internet. This method, though, does not allow access by a user who is out of range of the control link. It does allow a large bandwidth and therefore a high data rate. It is generally more expensive than land-line remote connections, but it may not be as reliable as the PSTN. In the USA, FCC rules dictate that any such radio control link be placed on the 1.25-m or shorter-wavelength bands.

Other Kinds of Signals on Control Links

Command, control and the desired communications data may be supplemented on control links by the presence of *telemetry*: the reporting of remotely made measurements of system parameters. Those parameters may include received signal strength, transmitted power, VSWR, heat-sink temperature, and so forth. Sometimes telemetry is sent continually to the control point and sometimes it is sent only on demand.

Occasionally, data multiplexing arrangements are used to send and receive digital data that are the desired communications signals. For example, a radioteletype (RTTY) modem may be remotely located along with the transceiver to which it is connected; the data to and from the modem is passed over the control link along with control and command data. Digitized audio may also be multiplexed into the data stream. More than one serial data stream may be combined in this way. One segment of the stream might contain digitized audio; another, the command and control data, audio or other desired transceiver signals; another, commands and telemetry from a remotely located amplifier; and still another, the commands and telemetry to and from a remote antenna rotator.

During command and control operations, a remote control system should have some means of verifying that commands were executed correctly and in a timely fashion. This is usually done by arranging for the unit being controlled to issue an acknowledgment of some kind after execution of the command. Such a handshake is required to establish positive control.

Legal Issues and Definitions

When an Amateur Radio station is being remotely controlled, it is said to be under *telecommand*.[64] The telecommand link, whether via radio or wire line, must be sufficient for the control operator to perform his or her duties. When the telecommand link is via Amateur Radio, the link radios are said to be in *auxiliary operation*.[65] An auxiliary station may be automatically controlled and may transmit one-way communications. Therefore, transmission of telemetry is not considered to violate the rule about codes and ciphers. When other telecommunications services are being used, the telecommand link is considered wire-line. In that case, encryption of remote-link data may be performed, subject to the conditions of the telecommunications service being used. Any amateur license holder may be the control operator, according to their individual privileges.

An amateur station under telecommand must make provision to limit transmission length to three minutes in the event of a failure of the control link. That rule explains why repeaters have "alligators" that cut you off before three minutes have elapsed. Reasonable steps must also be made to ensure that unauthorized persons cannot gain access to the unit to make illegal transmissions. A copy of the station license and the name, address, and phone number of the licensee must be posted at the station site. The same information for at least one designated control operator must also be posted.

A few questions arise about what location to report during a typical exchange via remotely controlled Amateur Radio. When the radio is in Flagstaff and the control operator is in Dayton, for example, where does the Ohio resident say he is? Well, he says he is in Ohio, of course, and adds that the radio is in northern Arizona. That kind of thing gets a lot of attention on the bands! During a contest, some clever person might be tempted to put a radio in each of several rare countries; he or she then would connect to each in turn to get multipliers. His or her efforts would be in vain, though, since those sorts of contacts are not usually counted in contest scoring.

Advanced Topics in DSP

A few advanced subjects are worth covering here. They represent exciting trends in DSP communications science that hold significant promise for the future.

Introduction to Adaptive Beamforming

Several articles and books, including this one, have discussed using adaptive filters as a way to build systems that either eliminate or enhance some time- or frequency-domain property of an input signal.[66] That is the basis for most embedded noise-reduction systems today. *Adaptive beamforming* extends the same techniques to the spatial domain. When it comes to antenna arrays, the goal is a certain radiation pattern that maximizes the received signal and eliminates unwanted signals. Arrays achieve their directivity by phasing of various elements. Placing element phase under DSP control opens some very interesting possibilities indeed.

In adaptive filtering, coefficients are adjusted on the fly according to the degree of correlation of the input data to past copies of itself. This correlation forces convergence to a set of filter coefficients that maximize (or minimize) output energy. Adaptive beamforming applies the idea of auto-correlation to the spatial distribution of radio signals.

Adaptive antenna arrays may be devised that automatically cohere; that is, they may be made to steer their patterns to produce signal cancellation (nulls) in the direction of undesired signals, even without knowledge beforehand of their direction of arrival. Algorithms have been invented that separate strong signals from weak signals when their directions of arrival are different. Other algorithms may distinguish among signals based on their distance from the array; a system may be constructed that is sensitive to nearby signals and insensitive to distant signals, or *vice versa*.

Still other signal characteristics may be used to steer an array. Transient signals may be rejected in favor of steady or *stationary signals*. The possibilities are virtually unlimited. Such systems are likely to be significant in the design of *software-defined* radios. To begin, we will restrict our discussion to arrays that null directional interference.

Take the simple example of a two-antenna array as in **Fig 12.1**. Two identical, omnidirectional antennas feed a summation network; one of the antennas is routed through an adaptive FIR filter, though, that conditions its signal. The unconditioned signal is called the *primary* signal and the other, the *reference* signal. Imagine that two radio signals are received: a desired signal and some interference. The desired signal and the interference are coming from two different directions. Both antennas receive both signals, but being at separate locations, their outputs are not precisely identical but are time-related functions of one another. Imagine a single signal arriving from azimuth angle θ. When the signal at the primary antenna is:

$$S_{PRI} = A \cos \omega_0 t \qquad (137)$$

the signal arrives at the reference antenna earlier by an amount of time:

$$\delta = \frac{l \sin \theta}{c} \qquad (138)$$

where l is the antenna separation and c is the propagation speed. The signal at the reference antenna is therefore:

$$S_{REF} = A \cos (\omega_0 t + \omega_0 \delta) \qquad (139)$$

The radiation pattern of the array may now be steered by altering the filter coefficients. Time of arrival is translated into direction of arrival by virtue of the array's architecture.

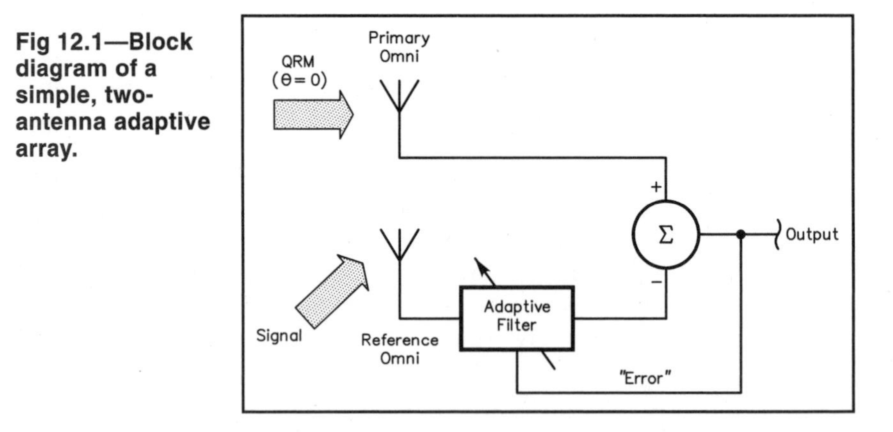

Fig 12.1—Block diagram of a simple, two-antenna adaptive array.

Upon convergence, the conditioned reference signal contains an interference component that is very nearly a match to the interference component of the primary signal and the interference is nulled in the summation. The system's output contains little interference, therefore, but still contains the desired signal, just as described in the case of adaptive noise reduction. This arrangement was originally studied by Howells in the late 1950s and later developed by Howells and Applebaum.[67]

In actual practice, primary and reference antennas usually feed separate receivers for amplification, selectivity and detection. The receivers add noise that may have significant impact on performance. Note that for this system to work properly, the interference must be strong compared to the desired signal, since the filter coefficients are determined almost exclusively by the interference. The desired signal will not be nulled if it is strong compared to the receiver noise. This set of conditions is commonly found in the field.

We are interested in the radiation pattern of the converged "side-lobe canceler" described above. It is pretty obviously bidirectional, since signals arriving from direction $-\theta$ produce the same situation. Widrow and Stearns (see Reference 56) have shown that the nulls it forms have depth proportional to the strength of the undesired signal. Just when the QRM gets stronger, the null improves! The pattern is always similar to a figure-eight, as shown in **Fig 12.2**.

Fig 12.2—Typical radiation pattern of a simple, two-antenna adaptive array.

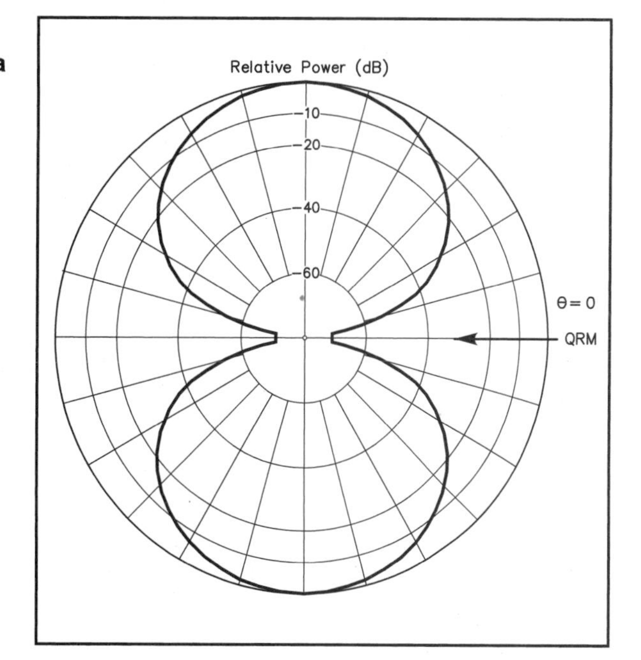

Under certain conditions, the desired signal may be so much stronger than the receiver noise that it tends to cancel itself. It is tempting to add artificial noise to the input to prevent this, but a better way to deal with this is the leaky LMS algorithm, discussed in Chapter 8. It is restated here in vector form:

$$\mathbf{h}_{k+1} = \gamma\, \mathbf{h}_k + 2\mu e_k\, \mathbf{x}_k \tag{140}$$

where **h** is the coefficients vector (the set of coefficients) and **x** is the input data vector. μ is chosen to be a positive constant less than unity. As mentioned before, values of μ greater than one may be tried to force the equivalent noise power downward, but the leaky LMS algorithm is only conditionally stable in this case.

Perhaps a more practical situation occurs in the presence of multiple interference sources. To cope with it, more than one reference omni must be used. Multiple sets of nulls may be formed in this way. See **Fig 12.3**. Both conditioned reference signals are subtracted in the final summation. In this case, each antenna is distance l from its neighbor, although spacings may be varied to achieve different goals. This system may produce two pairs of nulls, as shown in **Fig 12.4**. Note that each FIR adaptive filter may have as few as two coefficients (see Chapter 8).

If more than two interferers are expected, more reference antennas may be

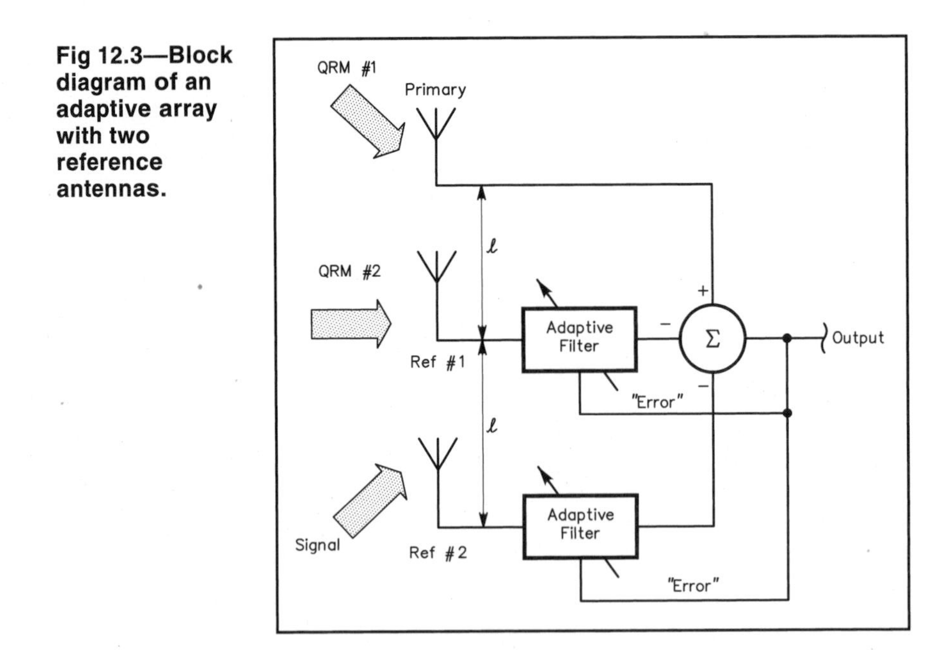

Fig 12.3—Block diagram of an adaptive array with two reference antennas.

Fig 12.4—Radiation pattern of a two-reference adaptive array when two QRM sources are present.

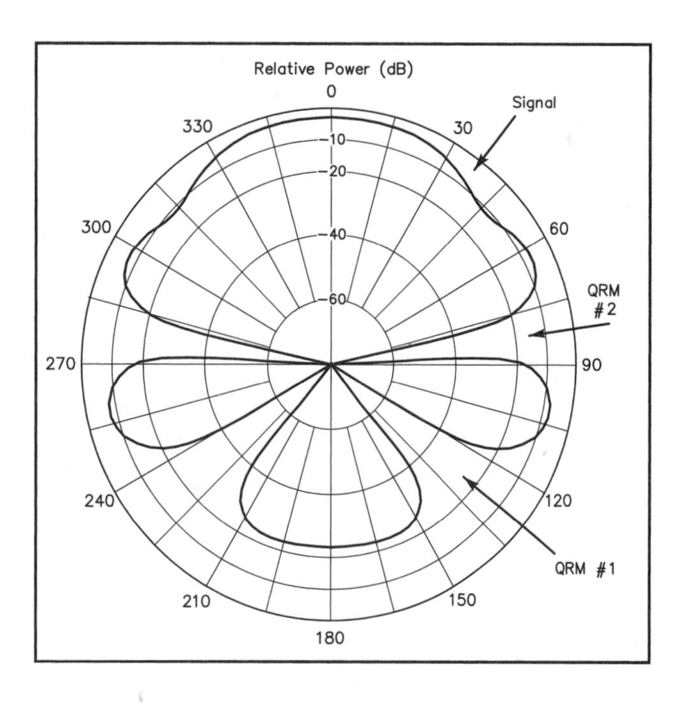

added. When the number of interferers exceeds the number of reference antennas, the system converges to the solution that minimizes output energy. That is a complex function of all input variables.

Manual Steering With a Pilot Signal

It is not hard to see that the pattern of an adaptive array may be manually steered by artificially placing a pilot signal in the direction of the desired null or lobe. In one type of adaptive beamformer, this manual control is retained along with the ability to adaptively null interference. The Howells-Applebaum side-lobe canceler is not preset to any particular look direction in the absence of signals; a pilot-signal adaptive beamformer relaxes to a predetermined directivity pattern in the absence of signals other than the pilot.

As array complexity grows, adaptive elements may shrink to have single weights (simple delay lines) to control radiation pattern. **Fig 12.5a** shows a six-omni array with $l=\lambda/2$ and Fig 12.5b, its radiation pattern. When fixed delays are inserted in the signal paths, the array may be steered to some extent, as shown in the example of Fig 12.5c. The main lobe is now centered on angle:

$$\theta = \sin^{-1}\left(\frac{\lambda_0 \, \omega_0 \, \delta}{2\pi l t}\right) = \sin^{-1}\left(\frac{c\delta}{l}\right) \tag{141}$$

Sensitivity is maximized at this angle because the incident wave produces

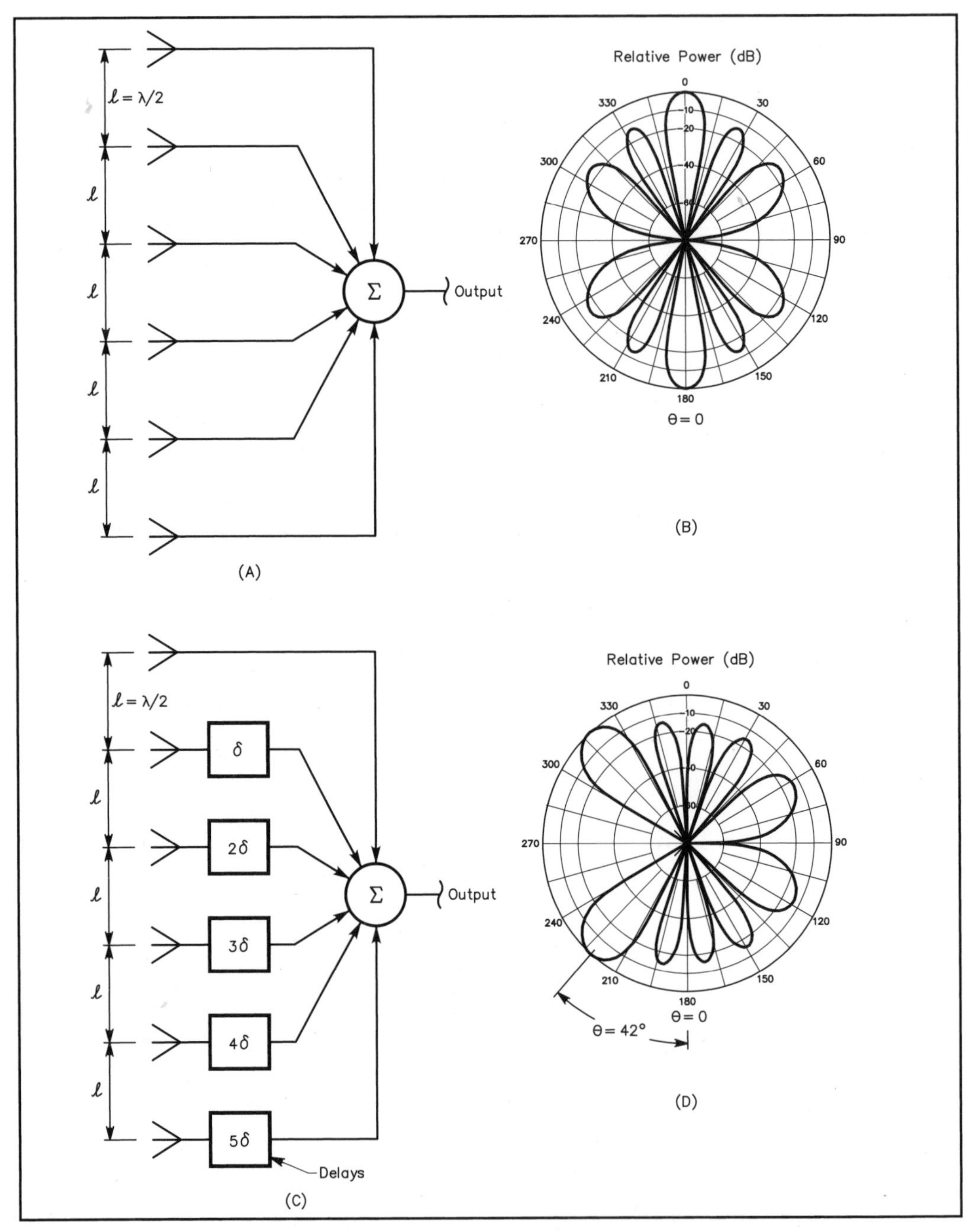

Fig 12.5a—A six-antenna linear array.
Fig 12.5b—Radiation pattern of (a) without delays in signal paths.

Fig 12.5c—The six-antenna array with delays inserted in signal paths.
Fig 12.5d—Radiation pattern of (c).

conditioned signals that are in phase with one another to add in the summation. As an example, when δ=250 ns and l=150 m, θ=30°. The system may be considered a pilot-steering system as the weights (delays) are manually set.

Spatial Architecture

To continue this line of thought, we must examine how antennas should be arranged to achieve some particular advantage in an adaptive beamformer. In practice, it may be found that placement does not matter much, since an adaptive system will take advantage of whatever spatial diversity it is given. Instinct would tell us that bigger is better. Still, an understanding of practical ways to control element phase is obviously needed and we have yet to examine narrow-band solutions to the adaptive beamformer problem.

One narrow-band solution presents itself in analytic-signal form, much as seen in the discussion of modulation in Chapter 5. Let us assume each antenna's signal is conditioned by a Hilbert transformer with weighting, as shown in **Fig 12.6**. Each signal is weighted by a complex factor:

$$H = A(\cos\phi + j\sin\phi) = Ae^{j\phi} \qquad (142)$$

where

$$\phi = \tan^{-1}\left(\frac{h_2}{h_1}\right) \qquad (143)$$

Now, any phase angle ϕ may be generated by adjusting the two coefficients h_1 and h_2. The absolute magnitude of throughput gain is just:

$$A = \sqrt{h_1^2 + h_2^2} \qquad (144)$$

We made the input signal narrow-band so that we could minimize the number of weights. Broadband signals could be handled by using a Hilbert transformer with many weights in each leg (a long, analytic FIR filter pair). It might even be possible to design analytic filters that control the phase of different signals separately within the passband.

The efficacy of the pilot-signal system extends only to the similarity of the pilot signal and the desired signal. That is, the pilot signal injected at the receiver site is designed to have characteristics that resemble those of the desired signal. Further, the presence of a pilot signal in the array output may render the output unusable. This drawback has led to the development of algorithms that switch the pilot signal on only when adaptation is performed, then the coefficients are frozen and the pilot signal switched off for normal operation.

The Griffiths Beamformer

L. J. Griffiths' algorithm is a take-off on the LMS algorithm.[68] It may be used with advantage where knowledge exists beforehand about the correlation

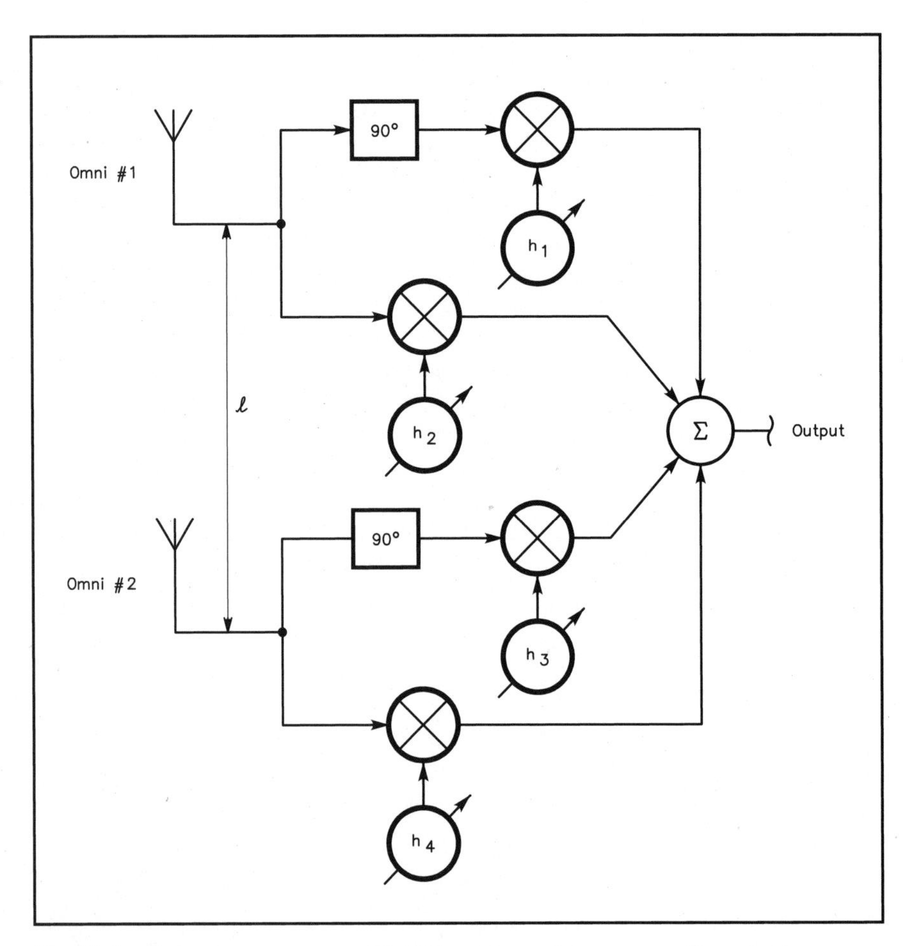

Fig 12.6—An antenna array with Hilbert transformers as the adaptive elements.

between the desired response, d_k, and the filter coefficients h_k. To do it, we have to re-formulate the LMS algorithm a bit:

$$\begin{aligned} h_{k+1} &= h_k + 2\mu e_k x_k \\ &= h_k + 2\mu\left(d_k - y_k\right)x_k \\ &= h_k + 2\mu d_k x_k - 2\mu y_k x_k \end{aligned} \qquad (145)$$

Now substitute the average value $E[d_k x_k]=S$ for its instantaneous value above and the result is Griffith's:

$$h_{k+1} = h_k + 2\mu(S - y_k x_k) \qquad (146)$$

Now S is fixed and the thing operates without a desired response input, d_k. It converges on the least-mean-squares solution like the LMS algorithm, but to put it to work, one must have knowledge of the desired signal's direction of arrival, its autocorrelation function, and the array geometry. As against that, the pilot-signal algorithm does not need to know the look direction or array geometry since the pilot signal could be transmitted remotely.

Frost's Adaptive Beamformer

Both the above-described systems use "fuzzy reasoning" to place restraints on performance so they don't go wild. They are useful in imposing predetermined conditions on that performance, so weak desired signals can do little to alter the patterns they produce. In the design of O. L. Frost,[69] a hard constraint is placed on the look direction. With this restraint, the sensitivity in the look direction is fixed without regard to the strength of desired signals from this direction.

A block diagram of a Frost beamformer is shown in **Fig 12.7a**. Fixed steering delays are once again used ahead of tapped delay lines. Now this is shown to be equivalent to the system of Fig 12.7b where the summation is carried out

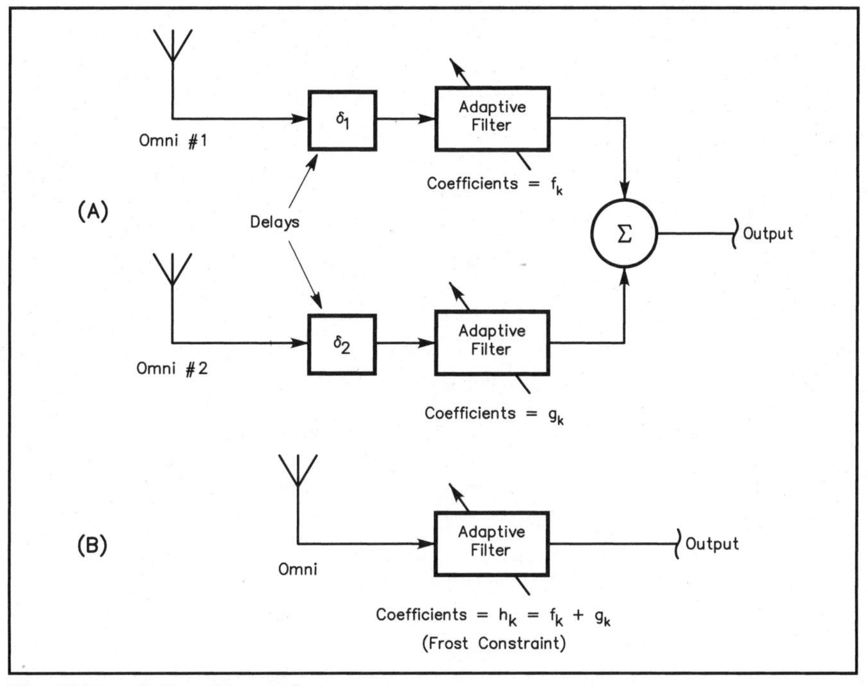

Fig 12.7a—A Frost beamformer.
Fig 12.7b—Equivalent circuit of (a).

as if the entire antenna processing array were a single processor. Each of the weights in this single processor is just the sum of the corresponding weights of each individual processor at Fig 12.7a. After the fixed weights of this single processor are found for a specified look direction, the weights of the adaptive processors may be varied as long as the *Frost constraint* is maintained. This way, the system obeys a fixed transfer function and filters the desired signals arriving from the look direction. It also minimizes output power by adapting itself to eliminate interference from other directions.

In the absence of the Frost constraint, and were the output power minimized, all the weights would go to zero and the output would disappear. The Frost constraint forces linear combinations of weights to be equal to certain constants. This causes the number of degrees of freedom to be less than the number of adaptive weights by the number of constraints placed. Degrees of freedom, in this case, is related to the number of adaptive nulls that may be maintained.

Distortion in Adaptive Beamformers and Super-Resolution

We may define distortions of both the desired signal and of the radiation pattern. As for the signal, distortion is produced by rapid variations in the adaptive weights and this is to be minimized in the steady state. In many cases, noise in the weights can have an effect on the output. While this seems contrary to information already given about DSP filters, we are now discussing non-Wiener behavior. Partial interference cancellation is a normal result and ideal characteristics sought are seldom achieved exactly.

Steering constraints have a large effect on output. If one of the conditions is that the desired signal be narrow-band, then the worst interference that might be generated would also be narrow-band. On the other hand, if the desired signal is broadband and the QRM sinusoidal, adaptive algorithms will try to modulate the QRM so that it cancels some of the desired signal at the QRM's frequency. These are effects with which every user of LMS-algorithm noise reduction is familiar.

Steady-state radiation patterns seem to depend as much on adaptation constants as input-signal fluctuation. Angular resolution of a regular antenna array is limited by the well-known Rayleigh criterion[70] for diffraction, also expressed by Fresnel in another form. A first cut at the 3-dB beamwidth of an array is:

$$\text{Beamwidth} = \frac{\lambda}{d}\,\text{radians} \qquad (147)$$

where λ is wavelength and d is the aperture diameter. When a signal is received with a high SNR, further improvement in resolution is possible through a concept developed by W. F. Gabriel in the late 1970s.[71]

Since antenna nulls are always sharper than lobes, bearings may be obtained more accurately by seeking a null. Radio fox hunters have known this fact for a long time. Exact information about direction of arrival may be used to turn a sharp

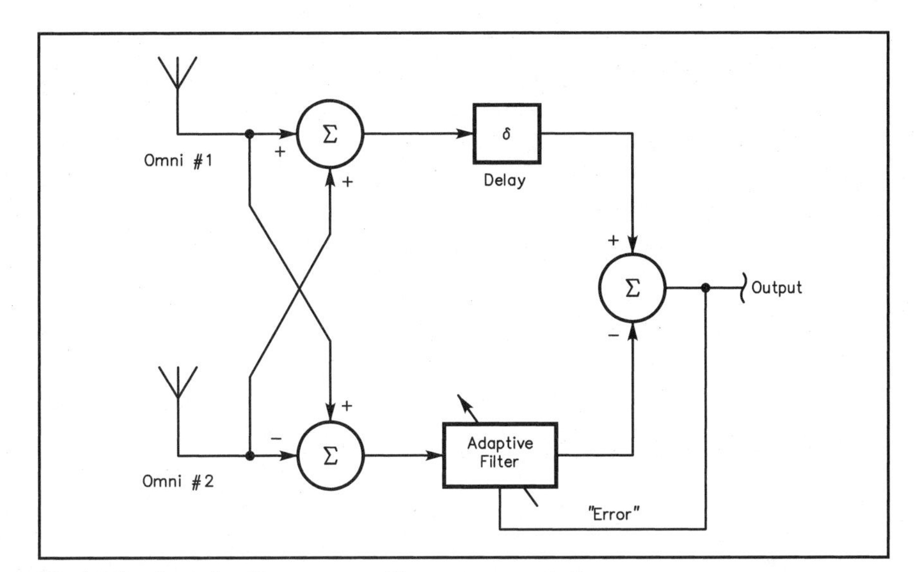

Fig 12.8—An adaptive array with super-resolution.

null into a sharp beam. The nulled output is simply subtracted from the signal received on a separate omni antenna, as shown in **Fig 12.8**, with weighting. The result is super-resolution. This system has seen little use in amateur circles, but it is expected to be a major part of software-defined radios that optimize performance in various increasingly hostile RF environments.

Software-Defined Radio Systems

A *software-defined radio*, or *SDR*, is one that is capable of operation on many frequencies, using many different modulation formats and bandwidths. It follows that most–if not all–of the modulation and demodulation activities are performed using DSP. An SDR employs hardware that is so universal that almost any possible radio function may be implemented by altering software. Those definitions carry with them certain implications about how SDRs must be designed to avoid excluding signal formats that have not been dreamed of yet.

The ideal SDR embodies something approaching a direct digital conversion scheme, as outlined in Chapter 9, and enough processing power to handle bandwidths and sampling rates that encompass all modulation formats envisioned. Examples of that are currently found in cellular telephones. The countries of the world have adopted different and incompatible standards for analog and digital speech transmission and reception. Your American cell phone will not necessarily work in South Africa, for instance. At the time of this writing, phones that comply with the various modulation standards in different countries are becoming available. They use hardware that allows them to cope with those different modulation standards and protocols; their functionality depends solely on how they are programmed.

You may be forced to push a button that selects what country you are in; but the ultimate potential of SDRs is realized only when they automatically determine what format to use. Principles of adaptive DSP and *automatic link establishment* (*ALE*) must obviously come into play. Adaptive beamforming provides one example of how DSP makes those kinds of magic things happen. Adaptive DSP and ALE are discussed below within the context of SDRs.

ALE for SDRs

ALE is an adaptive system whose goal is to determine the best channel on which to pass information. At the lowest level of ALE architecture is a receiver's frequency-scanning ability. A receiver may scan many frequencies in a relatively short period, listening for signals of interest and examining them when appropriate. If we are looking for the best frequency for communications, we had better have a large range from which to choose, or the exercise would be pointless.

The next-higher level of ALE incorporates some form of *selective calling*. That is because we need some way of discriminating among many possible signals on the air. The ham community has pioneered many digital transmission modes currently used with radio; we can easily find examples of selective calling among them. Almost every digital mode you can name supports selective transmissions, including AMTOR, packet, PACTOR, PACTOR II, CLOVER, GTOR and cellular telephones. In fact, what transmissions other than news bulletins and CW practice are not selective?

At ALE level three, we integrate the first two levels into a *path-quality evaluation (PQE)* scheme. When not passing traffic, all stations in the network are scanning assigned frequencies. Selective-calling tools are then used to attempt contact with a desired station on the available frequencies, one by one. Connections with the desired station results in some indication that we have been heard. At this point, scanning is suspended and some exchange of data is made to evaluate link quality. That may be as simple as a measurement of signal strength; or it may involve a measurement of the *bit-error rate* (*BER*), or both.

This procedure is known as *polling*. After all frequencies have been assessed, a decision about the best frequency to use is made. Either station may make this decision; but ultimately, one will call the other again on the best frequency to establish communications. In most such systems, PQE is made at the time when a connection is desired. After several exchanges are made, connectivity matrices may be stored for future use.

An ALE system involving many client stations that communicate with a single, central station may be designed so that the central station has the final say-so about what channel to use. That decision may be made with regard to what channels are already in use, as well as signal strength and BER.

Another method of determining connectivity is called *sounding*. In that procedure, stations that are not passing traffic periodically broadcast messages intended for the general consumption of all other stations in the network. Acknowledgment of sounding transmissions is not expected. Those stations that happen to be listening on a particular frequency evaluate the quality of reception, noting it. When it is time to connect to a specific station, therefore, each station has a better idea of which frequency is best.

While ALE has not caught on in Amateur Radio circles very much, public service and safety agencies became very interested in it beginning in the early 1980s. That was a time when fear of the "Evil Empire" was approaching another peak. Reasoning went along the lines that the first thing the enemy would do is shoot down or otherwise disable all our satellite and land-line communications, power sources, and so forth. We would be left with HF as the sole means of long-distance communications, as we would have been had lots of wire not been strung across the world following the invention of the telephone.

FED-STD-1045 and its higher-numbered cousins were invented and implemented to address those concerns.[72] An executive order from President Reagan got the whole thing rolling. Now those standards, which used a very robust modulation format primarily designed for HF, do not necessarily closely resemble what we want for SDRs; but the concepts they included remain valid to this day.

Adaptive DSP for SDRs

Having found some signal or signals that are likely candidates for communications, the next thing is to decide what modulation they use and whether it is something we are looking for. Spectral analysis via FFTs, auto-correlation and adaptive filtering may be valuable tools when it comes to determining what we are listening to. Certainly, analytic signals in DSP allow a flexibility that is unlikely to be surpassed in the near future. Only so many ways of modulating an RF signal are possible: Amplitude, phase and duration are the only ways to characterize signals that we know of. Those properties may be stated in different terms, of course, such as signal strength, frequency and bandwidth.

A DSP system may analyze a received signal in every conceivable domain because the signal may be exactly described in those terms stated above. That is not to say, though, that the information the signal contains may always be recovered easily. In actual practice, constraints have to be placed on adaptive SDR systems so they have a realistic chance of finding out what is going on before the signal disappears. Those constraints are determined by the system designer. One may assume that a particular signal appearing at one time is likely to reappear at some later time, and with the same traits it had the first time. That assumption makes an SDR designer's life a lot easier. For Amateur Radio, encryption of a signal is not allowed and modulation formats have to be publicly documented.

What Can SDRs Do For You?

That is a legitimate question, considering that SDRs have thrown a sort of monkey wrench into the works: How do we make sure SDRs behave according to regulatory guidelines set by the FCC and other bodies? In 2000, the FCC initiated requests for comments about its Notice of Proposed Rule-Making (NPRM) that sought input from the amateur community, among others, about how to deal with SDRs. Some activity centered on the fact that the Amateur Radio Service has played a prominent role in advancing the state of radio art over the years and that our service is a logical place to experiment with SDRs.

In fact, several Amateur Radio transceivers come close to being defined as SDRs. Some include the capability to completely update their firmware, either via a plug-in card or by download via the Internet. Some of us have been experimenting with SDRs without even knowing it! That capability certainly adds value to a transceiver, since it means, to some extent, that the unit will never be obsolete as long as someone is still willing to write code. Benefits of the SDR approach arrive in proportion to the accessibility of code and the number of programmers available to do the coding.

Manufacturers have found little benefit in cutting loose copies of their source code, though, especially since they obtained it at great expense. They want to keep whatever edges they have over the competition. Additionally, platforms for embedded DSP systems are based on many different microprocessors, which makes code portability a difficult goal. Perhaps that situation will change in the 21st century, but it is doubtful at best.

We have seen the success of Linux and similar open software systems. We have to ask, though, "How can you make a living by giving away your product?" Well, the answer is that you cannot. Everyone has to engage in a legal livelihood that provides for food, shelter and happiness; however, that does not mean you cannot share what you know with others and still flourish. Chances are excellent that you will learn at least as much as you teach. How can you lose?

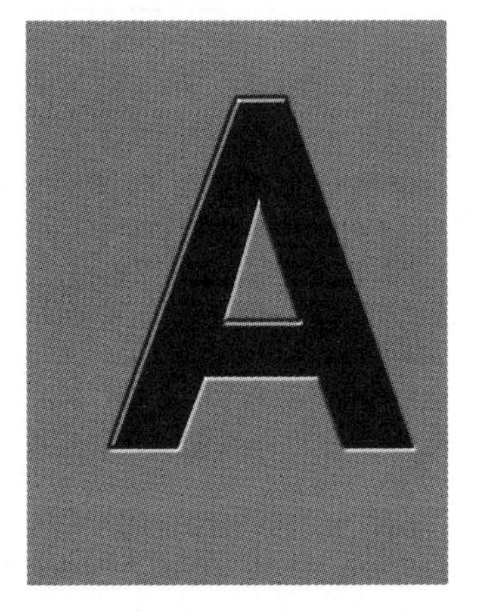

Conversion Loss of Passive, Commutating Mixers and the Mean Squares Method of Harmonic Analysis

The goal of LO drive to a double-balanced mixer is to alternately switch diode pairs in the quad on and off at the LO frequency. In the limiting case, LO drive is a square wave. Ideally, no diode is ever in a half-on or half-off state. This is equivalent to multiplying the RF signal by either 1 or –1. The spectrum of the product of such a multiplication, as in every other case, is dependent on the spectra of the inputs. To show why the conversion loss of an ideal, commutating mixer is about 3.9 dB, we analyze the case where the RF input is a sine wave and the LO is a square wave. See **Fig A1**.

The square-wave LO obviously contains a fundamental component and some harmonics. Conversion loss is defined with respect to the mixer output products corresponding to the fundamental of the LO. Using a technique called the mean squares method, the amplitude of the fundamental component of a square wave may be found. The amplitudes of the desired mixing products may thereby be found using that fundamental LO amplitude; hence, the mixer's conversion loss.

If some repetitive function of time f_t is known to be a component of another function of time g_t, and its relative phase is known, the amplitude of

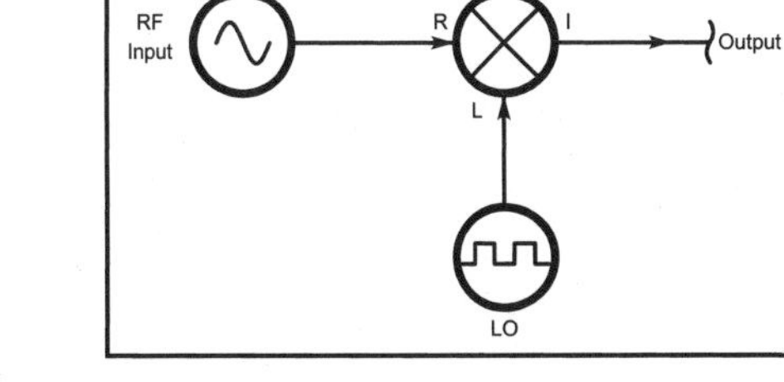

Fig A1—A commutating mixer with square-wave LO.

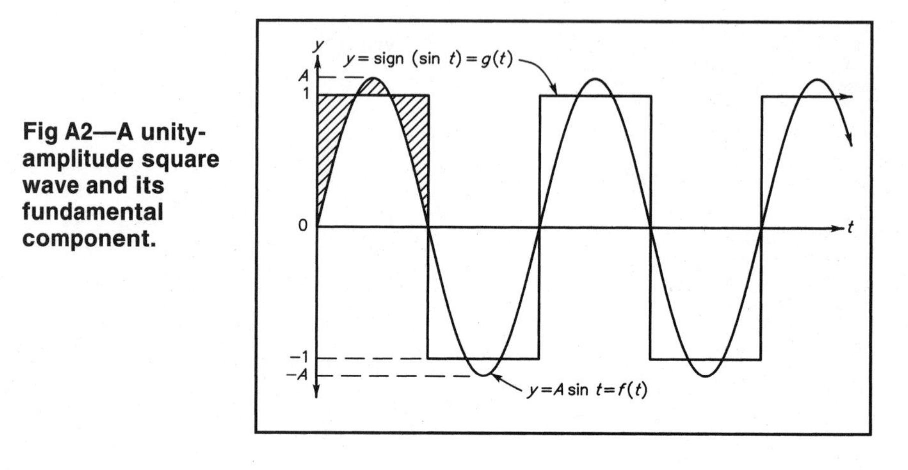

Fig A2—A unity-amplitude square wave and its fundamental component.

function f_t can be found using the mean squares method. In this case, we have g_t=sign(sin x), a square wave of unity amplitude; and f_t=A sin x, its fundamental component. See **Fig A2**.

Were g_t a waveform delivering power to a load, the power delivered would be equal to the sum of the RMS value of fundamental component, f_t, squared and the RMS value of the difference $g_t–f_t$, squared. In other words, the difference of the mean squares of g_t and f_t is equal to the mean square of the difference $g_t–f_t$. Here is how it works.

Find the mean square (RMS^2) value of g_t, the square wave. Then find the mean square value of $g_t–f_t$. Subtract from the mean square of g_t and solve for A, the amplitude of f_t. The mean square value of our square wave is 1. The peak amplitude of sine wave f_t is A and its mean square value is $A^2/2$.

Referring to Fig A2, the function $g_t–f_t$ represents the shaded area between the curves; it is equal to 1-A sin x. To find the mean square value of this function, find the average value of the square of the function over one half cycle. That may be written as:

$$
\begin{aligned}
MS &= \frac{1}{\pi}\int_0^\pi (1-A\sin x)^2\,dx \\
&= \frac{1}{\pi}\int_0^\pi \left(1-2A\sin x+A^2\sin^2 x\right)dx \qquad \text{(A1)} \\
&= \frac{1}{\pi}\left[x+2A\cos x+A^2\left(\frac{x-\sin x\,\cos x}{2}\right)\right]_0^\pi \\
&= \frac{A^2}{2}-\frac{4A}{\pi}+1
\end{aligned}
$$

We may solve for A by writing:

$$MS_{g_t - f_t} + MS_{f_t} = MS_{g_t}$$

$$\left(\frac{A^2}{2} - \frac{4A}{\pi} + 1\right) = 1$$

$$A^2 - \frac{4A}{\pi} = 0$$

$$A\left(A - \frac{4}{\pi}\right) = 0$$

$$A = \frac{4}{\pi}$$

The fundamental's peak amplitude is $4/\pi$, or greater than unity. To find out what happens in a commutating mixer, we use this LO amplitude in the basic mixing equation, Eq 44 from Chapter 5. In what follows, A is the amplitude of the RF signal being mixed:

$$y = \left(\frac{4\cos\omega_0 t}{\pi}\right)(A\cos\omega t)$$

$$= \frac{2A}{\pi}\left[\cos(\omega_0 + \omega)t + \cos(\omega_0 - \omega)t\right]$$

and the sum and difference products appear at amplitude $2/\pi$ relative to the RF input amplitude. Conversion loss is therefore:

$$-20\log\left(\frac{2}{\pi}\right) \approx 3.9\text{dB}$$

RMS Value of Sawtooth Waves

Calculation of the RMS value of a sawtooth or triangular wave is useful in analysis of distortion produced by look-up tables, either with or without interpolation. It has direct bearing on distortion at the output of DDS hardware and software, as described in Chapter 7. The following is another demonstration of how integral calculus finds the answer.

Take the repetitive sawtooth wave shown in **Fig B1**. Its peak amplitude is A and its period is b. Appendix A showed that the RMS value of a wave is the square root of the average value of the function's square taken over its period. To find the sawtooth's RMS value, we first have to write an equation that exactly describes its shape. In this instance, that is relatively easy: Over a single period, that shape is a straight line.

A line passing through the origin is defined by its slope alone. A line's slope is equal to how much it rises (or falls) for each increment in the other

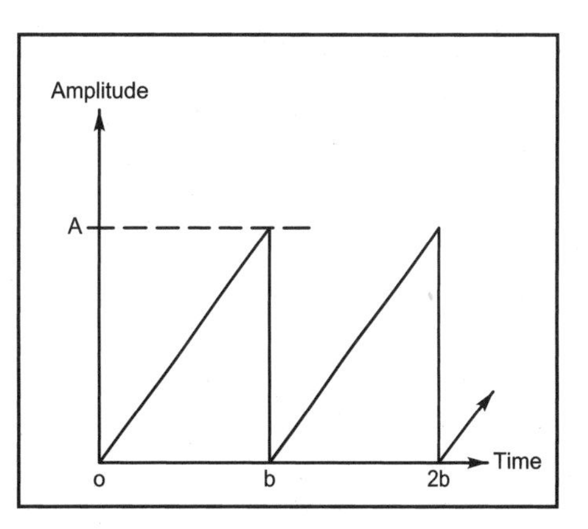

Fig B1—Sawtooth wave with peak amplitude A and period b.

dimension. We have already defined those quantities as A and b, so the slope of the line on which the sawtooth falls is simply A/b. The equation for the line may thus be written as y=Ax/b. Integrating the square of this over a single cycle, taking its average by dividing by b, and then taking the square root solves for its RMS value:

$$MS = \frac{1}{b}\int_0^b \left(\frac{Ax}{b}\right)^2 dx$$

$$= \frac{A^2}{b^3}\left[\frac{x^3}{3}\right]_0^b$$

$$= \frac{A^2}{3}$$

$$RMS = \sqrt{MS} = \frac{A}{\sqrt{3}}$$

Alternate Methods of SSB Demodulation

Methods for SSB demodulation described in Chapter 5 are certainly prevalent in amateur and commercial equipment, but the others should be mentioned: the Weaver method and the filter method. The Weaver method has at least one significant advantage over other methods and a few disadvantages. The filter method is not often used in DSP implementations and a brief discussion of it reveals why not.

Weaver Method

SSB demodulators shown in Chapter 5 use complex mixers to translate what would be the carrier frequency to a frequency of zero. Decimation by four is followed by an analytic BPF pair that determines the final selectivity. Analytic signal I+jQ, containing only positive frequencies, is the USB; signal I–jQ, containing only negative frequencies, is the LSB.

In the Weaver method, *the center of the desired passband* is mixed to zero frequency. Two identical low-pass filters are used to determine final selectivity, obviating the need for analytic filter-pair generation. The filtered signal is then complex mixed upward or downward in frequency to its final position using an oscillator frequency of about half the wanted bandwidth. Upward translation is used for USB, downward for LSB.

Note that the Chapter 5 method produces a filtered analytic signal containing both upper and lower sidebands, while the Weaver method produces only the desired sideband. That means the bandwidth of the filtered analytic signal in the Weaver method is half that of the former method. The sampling rate at that point, therefore, may be reduced by an additional factor of two: that's the big advantage of Weaver.

Like the other method, it mixes the input signal with a complex oscillator at one-fourth the initial sampling frequency. IF shift may be implemented in Weaver by altering the second injection frequency and providing a corresponding shift in the receiver's LO.

The main disadvantage of Weaver is that it doesn't support ISB operation by itself. That disadvantage arises chiefly in sideband-diversity reception of AM signals (see Chapter 5). Another disadvantage is that any dc offset appearing in the post-mix analytic signal is translated to a tone in the center of the final passband. That effect may be avoided, though, with careful processing.

Filter Method

The filter method of SSB is quite common in traditional, analog designs, but it is not often used in DSP because either of the above two methods brings advantages that the filter method lacks. The method is covered very well in the literature. Factors relating to its disuse in DSP are discussed briefly here.

The advantages of the I/Q methods above over the filter method include the ability to compute the SSB envelope prior to demodulation and the use of a single, convenient BFO frequency. The filter method does not necessarily involve complex signals, which may make it seem inherently simpler. It requires, though, that the desired passband be mixed to baseband in its correct sense; that is, with BFO injection on one side of the carrier frequency for USB and on the other side for LSB. That means either using two BFO frequencies in the analog sections of an IF-DSP receiver, or using a digital BFO that is not an integral sub-multiple of the sampling frequency. Alternatively, the sampling rate could be changed with sideband selection; but that presents difficulties in filter generation.

Those things and a few others have convinced DSP designers to stick with analytic-signal techniques. Complex signals lend themselves to other modulation types, too; thus, the uniformity of an analytic-signal approach is very attractive.

Details of FIR Filter Design

The Fourier-transform method for FIR filters is discussed here. Windowing techniques are included.

Fourier-Transform Method

As mentioned in Chapter 4, the basic idea behind the FT method is to sample a desired frequency response and then to find the inverse Fourier transform of that response. If the desired frequency response is H_n, where n is frequency normalized to the sampling frequency, then the impulse response of an FIR filter producing that frequency response ought to be:

$$h_k = \frac{1}{L}\sum_{n=0}^{L-1} H_n e^{\frac{j2\pi nk}{L}} \quad \text{(D1)}$$

for L values of k. That is true in general, but we can simplify matters quite a bit by taking a special case of that equation and reducing it to practice.

Consider the case of a low-pass filter having a brick-wall response, such as that discussed in the main text on page 2-14. H_n is a rectangular window of length L. Over all values of n, $H_n = 1$ inside the passband and $H_n = 0$ in the stopband. That fact simplifies the equation above, because the real part of the summation is just going to be the sum of some sinusoids of constant amplitude. For a finite value of L, we are not going to get a brick-wall response; we are going to get the kind of response shown at the top of **Fig 4.8** in the main text.

Let us adopt a cutoff frequency, f_0, normalized to the sampling frequency f_s. f_0 is then a fraction such that $f_0 < f_s/2$. The cutoff frequency cannot be greater than half the sampling frequency because of the aliasing phenomenon. Boiling down the equation, we may compute filter coefficients using:

$$h_n = \frac{\sin\left(\frac{2\pi n f_c}{f_s}\right)}{\pi n} \qquad \text{(D2)}$$

We now have a rectangular LPF.

Rectangular LPFs exhibit good transition-band characteristics, but do not have very good ultimate attenuation. As mentioned in Chapter 4, impulse windows may be applied to remedy the situation. A definite trade-off between transition bandwidth and ultimate attenuation exists.

Applying Windowing

To use the windowing technique, just multiply a window sequence with the rectangular filter's impulse response and take the product as the final filter's impulse response. For L values of k, we have:

$$h'_n = h_n w_n \qquad \text{(D3)}$$

Equations for window functions may be found in Reference 60.

References

1. G. B. Thomas, Jr., and R. L. Finney, *Calculus and Analytic Geometry*, 8th Ed., pp 303, 603, Addison-Wesley, Reading, MA, 1993.
2. R. M. Hutchins, Ed., *Great Books of the Western World*, 31st printing, vol 45, pp 163-251, Encyclopaedia Britannica, Chicago, IL, 1989.
3. P. Duhamel and M. Vetterli, "Fast Fourier Transforms: A Tutorial Review and a State of the Art," *The Digital Signal Processing Handbook*, V. K. Madisetti and D. B. Williams, eds., pp 7-1 through 7-42, CRC Press, Boca Raton, FL, 1998.
4. J. W. Cooley and J. W. Tukey, "An algorithm for the machine calculation of complex Fourier series," *Math. Comp.*, vol 19, pp 297-301, 1965.
5. U.S. Patent No. 2,275,735, R. T. Cloud.
6. T. Buxton, I. Hsu, and R. Barter, "Fetal electrocardiography," *Journal of the American Medical Association (JAMA)*, vol 185, pp 441-444, 1971.
7. A. V. Oppenheim and R. W. Schafer, *Digital Signal Processing*, pp 26-30, Prentice Hall, Englewood Cliffs, NJ, 1975.
8. W. E. Sabin and E. O. Schoenike, *Single Sideband Systems and Circuits*, 2nd Ed., pp 314-315, McGraw-Hill, New York, NY, 1995.
9. *Reference Data for Radio Engineers*, 6th Ed., p 23-7, Howard W. Sams, Indianapolis, IN, 1975.
10. T. Kalker, "On Multidimensional Sampling," *The Digital Signal Processing Handbook*, pp 4-1 through 4-17.
11. J. G. Proakis, C. M Rader, F. Ling, C. L. Nikias, *Advanced Digital Signal Processing*, pp 149-150, Macmillan, New York, NY, 1992.
12. K. Baudendistel, "DSP Implementation of Speech Processing," *The Digital Signal Processing Handbook*, pp 49-1 through 49-12.
13. L. R. Rabiner and B. Gold, *Theory and Application of Digital Signal Processing*, Prentice Hall, Englewood Cliffs, NJ, 1975.

14. T. W. Parks and C. S. Burrus, *Digital Filter Design*, John Wiley & Sons, New York, NY, 1987.
15. A. I. Zverev, *Handbook of Filter Synthesis*, John Wiley & Sons, New York, NY, 1967.
16. Op. cit. 7, pp 441-444.
17. Op. cit. 8, p 310.
18. Op. cit. 9, p 23-4.
19. F. Panter, *Modulation, Noise, and Spectral Analysis*, Chapter 5, McGraw-Hill, New York, NY, 1965.
20. D. Cobbe, KI6TN (SK), private communication, Escondido, CA, 1982.
21. R. C. Dixon, *Spread Spectrum Systems*, John Wiley & Sons, New York, NY, 1994.
22. P. Martinez, G3PLX, "PSK31: A New Radio-Teletype Mode," *QEX*, ARRL, Newington, CT, pp 3-9, Jul/Aug 1999.
23. W. E. Sabin, WØIYH, letter to the editor, *QEX*, Nov/Dec 1999.
24. W. B. Bruene, W5OLY, "MSK Simplified," Collins Radio Co. working paper WP-9262, 1971.
25. C. E. Shannon and W. Weaver, *The Mathematical Theory of Communication*, University of Illinois Press, 1963.
26. Op. cit. 9, p 41-23, Howard W. Sams, Indianapolis, IN, 1975.
27. B. Smith, "Instantaneous Companding of Quantized Signals," *Bell Systems Technical Journal*, Vol. 36, No. 3, pp 653-709, May 1957.
28. *Op cit 9*, Chapter 22.
29. J. Greefkes and K. Riemens, "Code Modulation with Digitally Controlled Companding for Speech Transmission," *Philips Tech Review*, 1970.
30. "Continuously Variable Slope Delta Modulation," MX-COM Application Note, *1997 Product Data Book*, Winston-Salem, NC.
31. L. Rabiner and R. Schafer, *Digital Processing of Speech Signals*, Prentice Hall, Upper Saddle River, NJ, 1978.
32. R. Hand, W8WQS, letter to the editor, *QEX/Communications Quarterly*, Jul/Aug 2000.
33. B. Moore, *An Introduction to the Psychology of Hearing*, Academic Press, London, 1989.
34. H. Fletcher, "Loudness, Masking and Their Relation to the Hearing Process and Problem of Noise Measurement," *Journal of the Acoustic Society of America*, Vol 45, 1969.
35. B. Scharf and D. Fishkin, "Binaural Summation of Loudness: Reconsidered," *Journal of Experimental Psychology*, Vol 86, 1970.
36. H. Fletcher, *Speech and Hearing in Communication*, ASA Edition, J. Allen, ed., American Institute of Physics, New York, 1995.
37. S. Stevens and H. Davis, *Hearing*, John Wiley & Sons, New York, 1938.
38. C. Harris, ed., *Handbook of Acoustic Measurements and Noise Con-*

trol, McGraw-Hill, New York, 1991.

39. J. Hall, "Auditory Psychophysics for Coding Applications," *The Digital Signal Processing Handbook.*
40. H. Fletcher and W. Munson, "Relation Between Loudness and Masking," *Journal of the Acoustic Society of America*, Vol 9, 1937.
41. B. Scharf, "Critical Bands," *Foundations of Modern Auditory Theory*, J. Tobias, ed., Academic Press, New York, 1970.
42. E. Zwicker, G. Flottorp and S. Stevens, "Critical Bandwidth in Loudness Summation," *Journal of the Acoustic Society of America*, Vol 29, 1957.
43. R. Schafer and L. Rabiner, "Design of Digital Filter Bands for Speech Analysis," *Bell System Technical Journal*, Vol 50, No 10, 1971.
44. R. Schafer, L. Rabiner and O. Herrmann, "FIR Digital Filter Bands for Speech Analysis," *Bell System Technical Journal*, Vol 54, No 3, 1975.
45. R. Stauffer, ed., *Charles Darwin's Natural Selection*, Cambridge University Press, London, 1987.
46. Patent pending.
47. US Patent No 3,349,184, Morgan, 1967; also see J. Flanagan and R. Golden, "Phase Vocoder," *Bell System Technical Journal*, Vol 45, No 9, 1966.
48. US Patent No 4,374,304, Flanagan, 1983; also see US Patent No 3,510,597, Williamson, 1970.
49. J. Ash, KB7ONG; F. Christiansen, KA6PNW; and R. Frohne, KL7NA, "DSP Voice Frequency Compandor for use in RF Communications," *QEX*, July 1994.
50. *Op cit 8*, Chapter 8.
51. F. Cercas, M. Tomlinson and A. Albuquerque, "Designing With Digital Frequency Synthesizers," *Proceedings of RF Expo East*, Intertec Publishing, Overland Park, KS, 1990.
52. W. Sabin and E. Schoenike, "Single Sideband Systems and Circuits," 1st ed., McGraw-Hill, New York, 1987, Chapter 9. This is the first known printed publication of a DDS-driven PLL.
53. G. Goldberg, "Frequency Synthesis Technology and Applications: A Review and Update," *QEX/Communications Quarterly*, Sep/Oct 2000.
54. U. Rohde, KA2WEU, "A High-Performance Fractional-N Synthesizer," *QEX*, Jul/Aug 1998.
55. M. Skolnik, *Introduction to Radar Systems*, McGraw-Hill, New York, 1962.
56. B. Widrow and S. Stearns, *Adaptive Signal Processing*, Prentice Hall, Englewood Cliffs, NJ, 1985.
57. B. Widrow and M. Hoff, Jr., "Adaptive Switching Circuits," *IRE WESCON Convention Records*, Pt 4, IRE, 1960.
58. C. Runge, *Z. Math. Physik*, Vol 48, 1903; also Vol 53, 1905.

59. B. Gold and C. Rader, *Digital Processing of Signals*, McGraw-Hill, New York, 1969.
60. M. Frerking, *Digital Signal Processing in Communications Systems*, Van Nostrand-Reinhold, New York, 1993.
61. A. Einstein, "On the Motion Required by the Molecular Kinetic Theory of Heat of Small Particles Suspended in a Stationary Liquid," *Annalen der Physik*, 1905.
62. M. Gruber, *ARRL Product Review Test Manual*, ARRL, 1992.
63. D. Reed, KD1CW, ed., *The ARRL Handbook for Radio Amateurs 2002*, 79th ed., ARRL, 2001, Chapter 14.
64. 47 CFR 97.213.
65. 47 CFR 97.201.
66. D. Smith, KF6DX, "Signals, Samples and Stuff: A DSP Tutorial, Pt 3," *QEX*, Jul/Aug 1998.
67. *Op cit 56*, Chapter 13.
68. L. Griffiths, "A simple adaptive algorithm for real-time processing in antenna arrays," *Proceedings IEEE*, Vol 57, 1969.
69. O. Frost III, "An algorithm for linearly constrained adaptive array processing," *Proceedings IEEE*, Vol 60, 1972.
70. Lord Rayleigh (J. W. Strutt), "On the theory of optical images, with special reference to the microscope," *Philosophical Magazine, Pt B*, Vol 42, No 5, 1896.
71. W. Gabriel, "Spectral analysis and adaptive array super-resolution techniques," *Proceedings IEEE*, Vol 68, 1980.
72. R. Adair and D. Peach, "A Federal Standard for HF Radio Automatic Link Establishment," *QEX*, Jan 1990; also see R. Adair, KA0CKS, *et al.*, "A Growing Family of Federal Standards for HF Radio Automatic Link Establishment (ALE)," Pts I-VI, *QEX*, Jul-Dec 1993.

About the ARRL

The seed for Amateur Radio was planted in the 1890s, when Guglielmo Marconi began his experiments in wireless telegraphy. Soon he was joined by dozens, then hundreds, of others who were enthusiastic about sending and receiving messages through the air—some with a commercial interest, but others solely out of a love for this new communications medium. The United States government began licensing Amateur Radio operators in 1912.

By 1914, there were thousands of Amateur Radio operators—hams—in the United States. Hiram Percy Maxim, a leading Hartford, Connecticut, inventor and industrialist saw the need for an organization to band together this fledgling group of radio experimenters. In May 1914 he founded the American Radio Relay League (ARRL) to meet that need.

Today ARRL, with approximately 170,000 members, is the largest organization of radio amateurs in the United States. The ARRL is a not-for-profit organization that:

- promotes interest in Amateur Radio communications and experimentation
- represents US radio amateurs in legislative matters, and
- maintains fraternalism and a high standard of conduct among Amateur Radio operators.

At ARRL headquarters in the Hartford suburb of Newington, the staff helps serve the needs of members. ARRL is also International Secretariat for the International Amateur Radio Union, which is made up of similar societies in 150 countries around the world.

ARRL publishes the monthly journal *QST*, as well as newsletters and many publications covering all aspects of Amateur Radio. Its headquarters station, W1AW, transmits bulletins of interest to radio amateurs and Morse code practice sessions. The ARRL also coordinates an extensive field organization, which includes volunteers who provide technical information for radio amateurs and public-service activities. In addition, ARRL represents US amateurs with the Federal Communications Commission and other government agencies in the US and abroad.

Membership in ARRL means much more than receiving *QST* each month. In addition to the services already described, ARRL offers membership services on a personal level, such as the ARRL Volunteer Examiner Coordinator Program and a QSL bureau.

Full ARRL membership (available only to licensed radio amateurs) gives you a voice in how the affairs of the organization are governed. ARRL policy is set by a Board of Directors (one from each of 15 Divisions). Each year, one-third of the ARRL Board of Directors stands for election by the full members they represent. The day-to-day operation of ARRL HQ is managed by an Executive Vice President and his staff.

No matter what aspect of Amateur Radio attracts you, ARRL membership is relevant and important. There would be no Amateur Radio as we know it today were it not for the ARRL. We would be happy to welcome you as a member! (An Amateur Radio license is not required for Associate Membership.) For more information about ARRL and answers to any questions you may have about Amateur Radio, write or call:

ARRL—The national association for Amateur Radio
225 Main Street
Newington CT 06111-1494
Voice: 860-594-0200
Fax: 860-594-0259
E-mail: **hq@arrl.org**
Internet: **www.arrl.org/**

Prospective new amateurs call (toll-free):
800-32-NEW HAM (800-326-3942)
You can also contact us via e-mail at **newham@arrl.org**
or check out *ARRLWeb* at **http://www.arrl.org/**

Index

Please use this form to give us your comments on this book and what you'd like to see in future editions, or e-mail us at **pubsfdbk@arrl.org** (publications feedback). If you use e-mail, please include your name, call, e-mail address and the book title, edition and printing in the body of your message. Also indicate whether or not you are an ARRL member.

Where did you purchase this book?
☐ From ARRL directly ☐ From an ARRL dealer

Is there a dealer who carries ARRL publications within:
☐ 5 miles ☐ 15 miles ☐ 30 miles of your location? ☐ Not sure.

License class:
☐ Novice ☐ Technician ☐ Technician Plus ☐ General ☐ Advanced ☐ Extra

---

Name ______________________ ARRL member? ☐ Yes ☐ No

Call Sign ______________

Daytime Phone ( ) ______________ Age ______________

Address ______________________________

City, State/Province, ZIP/Postal Code ______________________

If licensed, how long? ______________ E-mail ______________

Other hobbies ______________________

____

Occupation ______________________

For ARRL use only	DSP
Edition	1 2 3 4 5 6 7 8 9 10 11 12
Printing	2 3 4 5 6 7 8 9 10 11 12

From ______________________________

____

____

Please affix postage. Post Office will not deliver without postage.

EDITOR, DIGITAL SIGNAL PROCESSING TECHNOLOGY
ESSENTIALS OF THE COMMUNICATIONS REVOLUTION
ARRL—THE NATIONAL ASSOCIATION FOR AMATEUR RADIO
225 MAIN STREET
NEWINGTON CT 06111-1494

— — — — — — — — — — — please fold and tape — — — — — — — — — — — — — —